PHYSICS OF THE ENVIRONMENT

PHYSICS OF THE ENVIRONMENT

A. W. Brinkman

Durham University, UK

Imperial College Press

ICP

Published by

Imperial College Press
57 Shelton Street
Covent Garden
London WC2H 9HE

Distributed by

World Scientific Publishing Co. Pte. Ltd.
5 Toh Tuck Link, Singapore 596224
USA office: 27 Warren Street, Suite 401-402, Hackensack, NJ 07601
UK office: 57 Shelton Street, Covent Garden, London WC2H 9HE

British Library Cataloguing-in-Publication Data
A catalogue record for this book is available from the British Library.

PHYSICS OF THE ENVIRONMENT

ISBN-13 978-1-84816-179-5
ISBN-10 1-84816-179-4
ISBN-13 978-1-84816-180-1 (pbk)
ISBN-10 1-84816-180-8 (pbk)

Typeset by Stallion Press
Email: enquiries@stallionpress.com

Printed in Singapore.

To Jeannette
whose patience and support
has been beyond the call of duty

PREFACE

It is a truism to state that we are all subject to the environment within which we live. From it we derive the food and water that sustains life, the material and energy resources that makes that life more comfortable, and consign to it our waste products. Until very recently, these things were taken for granted. The resources seemed inexhaustible, while inexorable technological advances promised ever more comfortable lifestyles. Gradually, these perceptions have been changing under the irrefutable evidence of climate change.

The first intimations that all was not well came from the unexpected discovery of ozone depletion and the dramatic 'hole' over Antarctica. It soon became apparent that this was a direct result of human activity — the planet is not necessarily able to absorb all our waste. It is now widely appreciated that resources are indeed limited and that consumption, particularly in the developed world, might need to be reined in if others less fortunate are to be able to enjoy some of the benefits. However, it is the consequences of climate change that have engaged the news media the most and set political agendas, at least in Europe. While few among the scientific community dispute the twin observations of steadily increasing temperatures and rising levels of carbon dioxide, the actions required to reverse the trends are necessarily political and economic. Science alone cannot do this; it can name the dangers and even suggest possible solutions, but ultimately there are some decisions that are not within its province.

This book is a somewhat expanded version of a course on Environmental Science given by the author to second and third year Physics honours undergraduate students at Durham University. The course was intended to provide a bridge between the different individual specialisms that make up the general field of environmental science — one that would be accessible to the average student at this level of his or her degree programme. Too brief in many aspects to be in any sense comprehensive, the course did set out to introduce students to the diversity of the subject. The approach was meant to be rigorous and therefore to some degree mathematical, though with the hope that the physics would not be obscured by the mathematical 'trees'. The same philosophy has been adopted in 'Physics of the Environment' and the mathematical content has been kept to a minimum that is commensurate with a proper understanding of the subject.

There is in reality a dearth of physics textbooks that cover the breadth of material at a level appropriate for undergraduate students of physics.[a] Specialist texts are by definition 'specialist', focusing in depth on the particular field in question and are usually more suitable for postgraduate study. Often texts on environmental science are primarily concerned with ecological matters and on the whole do not address themselves to the physical sciences, or if they do, it is only at a superficial level. Some have clear polemic agendas.

With the exception of the final chapter, the rationale behind this book has been to develop the science and technology of the subject. That is not to say that there is an absence of material that some might deem controversial e.g. the inclusion of a section on nuclear power, a critique of the shortfalls of renewable energy, the so-called hydrogen economy all being cases in point. As part of the current and possible future environmental science landscape, such topics should not be ignored, but the intention has been to discuss the science in as neutral and unbiased manner as possible.

Finally, the book is by now long overdue and it would be unjust not to acknowledge the patience and forbearance of the editor, Katie Lydon, on this project.

A. W. Brinkman
Durham, December 2007

---

[a]Environmental Physics, 2nd edn by Egbert Boeker and Rienk van Grondelle, Wiley (1999) being an honourable exception — it was also the course text.

CONTENTS

Chapter 1

INTRODUCTION

1.1 Introduction

Narrowly defined, 'Physics of the Environment' simply refers to the body of physical laws governing the environment within which we live. In its wider usage, the term is understood to encompass a more active human involvement, including on the one hand the observation and study of the environment and on the other the employment of physical laws and technologies for (in principle) the betterment of all. This text attempts to include both aspects. No claim is made that the content is in any sense comprehensive, but the material has been chosen to reflect the main issues of the moment, current technologies and some of the options for the future.

People in agricultural communities in both developed and developing countries alike live lives that are closely tied to the environment. Work is dictated by the seasons and success or disaster by the vicissitudes of the weather — too much rain at the wrong time can spell disaster. In developed countries, infrastructure exists (e.g. insurance, government compensation schemes, etc.) to mitigate the effects, but nevertheless the connection is close and tangible.

In urban communities, whether it is a large metropolis in an industrialised country or an often equally large shanty town from a poor country, life seems much more detached from the natural environment. Livelihoods have to be earned using different specialised skills, often based in the crowded workplaces of the factory or the office. The concentration of resources in towns allows for the development of centralised facilities, hospitals, schools, markets, and government offices. Expensive to maintain and operate, such services can only be sustained if the catchment area is sufficiently populated, which is clearly not the case in rural districts. Of course, urban workers are no more immune to the vagaries of life than their counterparts in the village. Governments change and companies go bankrupt or get taken over, and employees are made redundant as a result.

The lives of city dwellers in both the first and third world countries, are also intimately dependent on the environment, even if it is not immediately apparent. The food has to be grown somewhere, waste has to be removed and safely disposed of, and electricity generated from fuel extracted in various ways from the environment. All societies, whatever their level of development, are ultimately and *absolutely* dependent on the environment for the provision of their needs, be it for food, energy or waste disposal.

A proper understanding of our dependence on the natural world allows us to use the available resources in sustainable ways. This is not merely a question of self-denial, but of the intelligent deployment of technology based on correct science for the greatest benefit of humankind.

1.2 Scientific Method

There are in essence three premises on which the scientific method is based. Firstly, we assume that observations made through our senses (and by extension scientific instrumentation) are an objective representation of reality. Secondly, we assume that the world may be described by a set of self-consistent rules that are constant in time and space. Thirdly, we assume that the observed phenomena have discernable causes, i.e. causality.

Measurement lies at the heart of observation. Quantification of the observable allows comparison and verification with observations made elsewhere and at other times. The human senses vary considerably from individual to individual, and even for the same individual may change over time and in response to circumstances. Visual observation, for example, depends strongly on the ambient light level as the iris adjusts the eye's aperture to suit light conditions. Evidently, there is a need for sophisticated instrumentation with a performance that is known and which can be calibrated against internationally agreed standards. In tandem with this, there must be agreed scales of units that have universally accepted definition. Only then, when measurements are repeated by others or with more precise instrumentation, can errors and uncertainties be corrected and resolved.

It is not possible to prove that the laws of nature are, as we perceive them on Earth, constant and apply throughout the universe. Nonetheless, it does seem to be the case. Frequently, our understanding of physical laws has to be refined as the 'frontiers' of knowledge and experience are increased. For instance, Newton's laws of motion should now be viewed as a low velocity limit, since at sufficiently high speed the relationships between time and space have to be described in relativistic terms. Similarly, when the dimensions of a system are very small, it must be described in quantum mechanical terms. Neither implies any conflict with 'normal' classical mechanics, only that there are dimensional regimes when one particular set of laws provides an adequate description. (In the latter case, classical mechanics may be thought of as the large scale approximation of quantum mechanics.)

In many situations, the principle of cause and effect may be demonstrated by laboratory experiments. A deliberate test can be set up to examine whether, if under a given stimulus, the system under study responds in the expected manner. This implies the existence of some idea or 'model' of the system's behaviour. In other words, having carried out a variety of quantified measurements, and having thought carefully about the process, some hypothesis has been developed to describe the observed behaviour. The hypothesis should have predictive power, i.e. if the system is excited in this way, then according to the hypothesis it ought to respond in that way. If it does not, then having ensured that there was no trivial mistake or error, the hypothesis has to be refined or discarded as necessary. Eventually, this process will reveal a working thesis that would appear to explain the results.

In the physical sciences, such models are generally expressed in mathematical terms that allow precise calculation and manipulation of observed values with a degree of rigour not always available in other branches of science. The expression of accepted models in mathematical form allows the extrapolation to situations beyond the laboratory into un-testable regimes. When harnessed to the processing power of modern computers, it becomes feasible to simulate, for example, the effects of different levels of atmospheric carbon dioxide on the rate of global warming and to determine concentrations that should not be exceeded.

1.3 Contents

This book is intended for use mainly by students of physics in their second or subsequent year of undergraduate study. It presumes a basic familiarity with the classical laws of Newtonian physics, an understanding of SI units and a corresponding level of mathematical competence. The text aims to show that the application of the laws of physics at a relatively elementary level can provide a sound description of the environment within which we live.

Often confused with climate, the daily weather is the result of vertical and horizontal movement of air. Broadly speaking, the former governs rain and precipitation, and the latter winds. Atmospheric motion takes place on the global scale, where interaction with the rotation of the Earth and the oceans give, for example, the Trade Winds, and dictates local averages of temperature and rainfall. The time scales are short in relation to those associated with changes of climate, although trends in the weather may well be a direct consequence. The dynamic behaviour of the atmosphere constitutes the subject matter of Chapter 2.

Global warming, the seemingly remorseless increase in the mean temperature of the planet, is probably the topic of greatest concern at the present time. The basic physical processes underpinning global warming, radiative forcing, feedback effects, greenhouse gases etc., form the content of the Chapter 3. The main culprit is generally accepted to be the emission of carbon dioxide from power stations and transport. A short account is included of the abortive attempts to restrict these emissions under the Framework Convention on Climate Control — firstly, the voluntary Rio de Janeiro proposals and the subsequent more binding (but equally unsuccessful) Kyoto Protocol.

Despite current concerns over climate change, fears about ozone depletion predate those of global warming and led to the adoption by virtually the entire world of the first, and to date the only successful international agreement on environmental controls: the Montreal Protocol. The role of ozone in screening out harmful ultraviolet radiation and the mechanisms behind its depletion are discussed in Chapter 4.

The basic mechanisms of heat transfer are reviewed in Chapter 5, which lays the foundations for subsequent chapters on power generation and transport.

The generation of electrical power from the combustion of fossil fuels is discussed at some length in Chapter 6. Embedded within this chapter is a review of the basic principles of thermodynamics that govern and limit thermal efficiency. The technologies for coal and gas fired generation and their relative merits are then described, including new techniques for the sequestration of CO_2 that could prove important in managing greenhouse gas emissions.

There are, of course, alternative sources of carbon-free power generation, namely nuclear and the renewable sources. The physics of nuclear power, discussed in detail in Chapter 7, is frequently ignored or glossed over in environmental textbooks, partly because the dynamics of neutrons within a reactor core and the problems of sustaining a chain reaction are challenging and complex. There is also probably a widespread assumption that the technology will always be too dangerous to be adopted (especially post-Chernobyl) except in a few countries, such as France, Japan and South Korea, which have elected to 'go nuclear' for reasons of national interests.

Renewable sources, which is really a collective term for a disparate group of technologies including wind, solar, biomass, hydroelectric and geothermal, pose particular problems of their own. The basic physics and technology of these energy sources are discussed in Chapter 8, together with some comments on the problems of their utilisation. With the exception of geothermal sources, renewables are 'land hungry' and variable (especially wind). Additionally, the power generated at a single site is low compared with a conventional power station making integration into grid-based networks difficult.

Chapter 9 is devoted to issues of transportation. One of the great freedoms of modern times has been the freedom to travel and it is one that has, in the industrialised countries at least, come to be taken for granted. The invention of the internal combustion engine coupled with the discovery of extensive reserves of cheap oil duly led to the widespread ownership and use of private cars. Unfortunately, they are a major source of CO_2 emission and despite significant advances in engine design and the use of catalytic converters, they will continue to be so as long as they burn hydrocarbon fuels. Potentially hydrogen may be used as a clean fuel for cars, either as a direct replacement for hydrocarbon fuels in adapted internal combustion engines, or in fuel cell-electric motor combinations. The technological challenges inherent in the 'hydrogen economy' (as it has come to be called) are demanding. Hydrogen is a gas at ordinary temperatures and pressures, making on-board storage in vehicles difficult. None of the currently available solutions have so far emerged as superior to its competitors. In addition, although it is one of the most abundant elements on Earth, hydrogen only exists in compound form (e.g. in water) and not as the free element required for its use as a fuel. It is a *secondary* source that still has to be produced from some other *primary* energy source.

All waste emissions, whether gaseous or liquid, must ultimately be accommodated within the environment. Harmful products have to be dispersed and diluted to safe levels — in the case of liquid effluents usually into rivers or the sea; in the case of smoke and gases, into the atmosphere. The mechanisms of dispersal, principally diffusion and advection, are discussed in Chapter 10. As a rule, individual systems are much too complex to be described by analytical solutions and invariably have to be modelled on a case-by-case basis. Nevertheless, they are physical processes governed by common general principles and the purpose here is to provide some intuitive appreciation of the underlying physics.

Groundwater flow, or the motion of water *within* the ground, is of considerable significance for human public health. Much of the water used for human consumption is derived from what amounts to vast natural underground reservoirs, known as aquifers. They are replenished by the same diffusion mechanisms that govern the dispersal of waste products into the environment, underlining the universality of these processes.

The final chapter considers some of the current ethical issues. It does not focus on the 'hard science' *per se* but on less tangible questions relating to the stewardship of the environment. The energy deficit that is currently on the minds of western governments, and which is engendering a renewed interest in nuclear power, is reviewed. How should planning authorities estimate risk and harmful issues and balance these against their potential benefits? How their does a democratic government reconcile legitimate competing and conflicting interests? A related topic is the consumption of finite material resources, arguably of greater significance in the long run than climate change. Ultimately, such questions often reduce to a matter of individual lifestyle choices and corporate decisions; in short, how we choose to steward the environment.

Chapter 2

STRUCTURE AND DYNAMICS OF THE ATMOSPHERE

2.1 Introduction

We are perhaps most aware of the impact of the atmosphere on our everyday lives through the weather. The physical parameters of temperature, rain and wind directly affect the way we organise our lives determining not only our physical needs in terms of clothing and heating, but also the way we feel including our perceptions of the environment. The weather itself is a consequence of the vertical and horizontal motions of air, evaporation and precipitation, radiative transfer etc. that take place in the atmosphere. These in turn are affected by the Earth's rotation and orbit about the Sun, and the proximity of significant geographical features. In this chapter, we shall endeavour to develop a physical understanding of the processes and apply them to aspects of daily weather. At this juncture, it is important to make the distinction between *climate* and *weather*. The latter is concerned with the immediate and essentially local variations of the variables of temperature, rainfall, sunshine, wind velocity and the like. The term climate implies some sort of average of these parameters over time and space and has a global character. We shall be discussing global climate and climate change in the next chapter.

In order to appreciate the dynamical physical mechanisms that occur within the atmosphere, it is first necessary to know something about its structure and composition. The latter changes with time and geographical location due to the uptake of water vapour or otherwise. The changes of phase in evaporation and condensation of water involve the exchange of latent heat and some understanding of the thermodynamics of these processes is essential. Wind is the physical displacement of air which has mass and hence inertia. We shall, therefore, consider the principal forces that give rise to wind and determine its speed and direction.

2.2 Structure and Composition of the Atmosphere

2.2.1 *Large scale vertical structure of the atmosphere*

Vertically, the atmosphere may be divided into four layers ('spheres') defined by the sign of the temperature gradient with altitude, separated by regions ('pauses') where the temperature is static as shown in Fig. 2.1. The troposphere, the lowest and densest part of the atmosphere, is characterised by a temperature that decreases steadily with altitude, and this is where virtually all the weather occurs. It extends

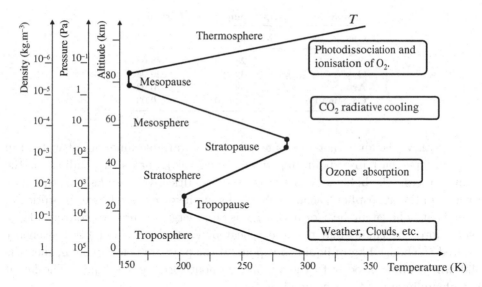

Fig. 2.1   Large scale structure of the atmosphere.

from sea level to an altitude of about 15 km, and is separated from the stratosphere by the tropopause.

The stratosphere, which extends over about 30 km, is the important ozone bearing layer which screens out harmful ultraviolet radiation from the Sun (see Chapter 4). The ozone photochemistry causes a reversal of the temperature gradient and a rise in temperature. Above the stratosphere further reversals in sign of the temperature gradient define in turn the mesosphere (negative temperature gradient) and the ionosphere, sometimes referred to as the thermosphere (positive temperature gradient), where the gas is fully ionised by absorption of energetic solar radiation.

It will be apparent from Fig. 2.1 that although the temperature gradient repeatedly changes sign, the pressure and density decrease exponentially with altitude. It is common practice in meteorology to use pressure as the measure of height above sea level, since it is the more important parameter for weather than altitude.

2.2.2 *Composition of the atmosphere*

Dry air is predominantly composed of nitrogen and oxygen, with small amounts of carbon dioxide and argon. In addition, there are trace amounts of methane (from the decomposition of vegetable matter), nitrous oxides (from both natural and anthropogenic sources), hydrogen and the other inert gases. The mean composition by weight of the main components is listed in Table 2.1. The relative concentrations of these gases with respect to nitrogen is roughly constant throughout the troposphere, stratosphere and mesosphere.

Table 2.1   Principal components of dry air.[1]

Gas	Molecular weight	Mass (%)
N_2	28.016	78.08
O_2	32.00	20.95
Ar	39.444	0.93
Carbon dioxide	44.010	0.03

In reality, the air is not dry but will contain variable amounts of water (up to several percent) in vapour, liquid and solid (ice) phases, depending on the circumstances. The behaviour of the weather is critically dependent on the mass of water in the atmosphere, as all changes of state involve the release/absorption of latent heat. The latent heat of fusion L_F is the energy required to convert 1 kg of a solid into the liquid, and the latent heat of vapourisation (L_V) is the energy required to change 1 kg of liquid to vapour. If the pressure is low enough, then the solid will sublime directly to the vapour, without forming the liquid. The latent heat of sublimation, L_S is being given by:

$$L_S = L_F + L_V. \tag{2.1}$$

Equal amounts of heat are released when the phase change is reversed.

A common measure of the moisture content is the mixing ratio. θ_w defined as the mass of water vapour per unit mass of dry air[a]:

$$\theta_w = \frac{m_w}{m - m_w} \tag{2.2}$$

where m_w is the mass of water vapour in mass m of air respectively. At a free surface, water will both evaporate and condense at rates determined principally by the temperature and partial pressure of water vapour (p_w) above the surface. If the water is contained in an enclosed volume, it will establish an equilibrium or saturation partial pressure (p_{sat}) where the rates of evaporation and condensation just balance. This gives rise to another widely used parameter, the relative humidity, R_H defined as the ratio of the observed mixing ratio to the saturation mixing ratio (θ_{sat}) i.e. the mixing ratio when the water vapour partial pressure is saturated:

$$R_H = \frac{\theta_w}{\theta_{sat}} \approx \frac{p_w}{p_{sat}}. \tag{2.3}$$

Relative humidity is normally expressed as a percentage and by definition, $R_H = 100\%$ implies that the moisture is condensing. In practice, droplets will tend to form on nucleating points (e.g. dust, surface asperities etc.) at lower relative humidity. It is common place, for example, to specify 95% relative humidity (non-condensing) as the maximum humidity rating for instrumentation.

---

[a]Although dimensionless, θ_w is normally expressed as kg/kg or gm/kg.

2.3 Vertical Motion of Air

2.3.1 *Hydrostatic equation and lapse rate*

In understanding the vertical motion of air in the atmosphere, we typically consider the behaviour of a small *parcel* of air of unit mass as it ascends through the troposphere. In doing so, the parcel will do work against gravity, gaining gravitational potential energy in the process. Initially, we shall assume that the air is dry and that there is no transfer of heat between the parcel and the surrounding atmosphere, $(\delta Q = 0)^2$, i.e. the adiabatic or constant entropy (isentropic) condition. The adiabatic case is frequently used as an ideal reference or 'limiting case' in many thermodynamic situations, however, it should be remembered that the real world is not ideal and there will always be some exchange of heat with the surroundings.

Suppose that the parcel of air is a slab of unit area and thickness dz located at an altitude z where the pressure is p and at the top of the slab $(z + dz)$, it is $p + dp$ as illustrated in Fig. 2.2. In the first instance, we will assume that the parcel is in static equilibrium, by which we mean that the net pressure force exactly balances the gravitational force.

$$dp = -g\rho_d dz \qquad (2.4)$$

where g is the acceleration due to gravity. Equation (2.4) is known as the *hydrostatic equation* or more properly the *equation of hydrostatic equilibrium*.

Further assume that the gases comprising of dry air follow the ideal gas laws which relate the pressure (p) and volume (V) of n moles of gas to the temperature (T). For a simple gas:

$$pV = nRT = \frac{m}{M}RT \qquad (2.5)$$

where R is the universal gas constant, M the molecular weight and m the mass. In a mixture of ideal gases, the partial pressure of the ith component can be written:

$$p_i = \frac{m_i}{V}\frac{R}{M_i}T = \rho_i R_i T \qquad (2.6)$$

where ρ_i and R_i are the density and specific gas constant $(\mathrm{J\,kg^{-1}\,K^{-1}})$ of the ith constituent respectively. The total pressure of the mixture (i.e. dry air) will be the

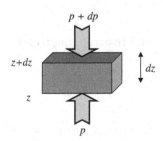

Fig. 2.2   Vertical forces on an elemental slab of air of thickness dz at an altitude z.

sum of the partial pressures:

$$p_{\text{Total}} = \sum_i p_i = T \sum_i \rho_i R_i = \rho_d R_d T \tag{2.7}$$

where R_d is the specific gas constant of dry air[3] ($287\,\text{kJ}\,\text{kg}^{-1}\,\text{K}^{-1}$). Substituting in the hydrostatic equation (2.4) and integrating:

$$\frac{\partial p}{p} = -\frac{g}{R_d T}\partial z, \quad \text{i.e. } p(z) = p(0)\exp\left(-\frac{z}{z_0}\right); \quad z_0 = \frac{R_d T}{g} \tag{2.8}$$

which is consistent with the observed exponential dependence of pressure on altitude (Fig. 2.1). Taking the mean temperature of the troposphere to be $250\,\text{K}$, $z_0 \approx 7.3\,\text{km}$ for dry air.

Lapse rate is defined as minus the rate of change of temperature with altitude (i.e. $-\partial T/\partial z$) and to derive this we start from the First Law of Thermodynamics[b] which relates the amount of heat added (δQ) to the temperature, pressure and volume of a substance, and for the adiabatic case may be written as:

$$\delta Q = c_V dT + p\,dV = 0 \tag{2.9}$$

where c_V is the specific heat at constant volume. As the parcel of air rises, it will experience a reduction in the ambient pressure and will therefore expand. It is adiabatic so[2]

$$pV^\gamma = \text{constant} \quad \text{and} \quad \therefore d(pV^\gamma) = V^\gamma dp + \gamma p V^{\gamma-1} dV = 0 \tag{2.10}$$

where γ is the ratio of the specific heats (c_p/c_V; c_p is the specific heat at constant pressure). From (2.10) we get

$$p\,dV = -\frac{V}{\gamma}dp. \tag{2.11}$$

Adding $(V/\gamma)dp$ to both sides of (2.9) and using (2.11) gives

$$\frac{V}{\gamma}dp = c_V dT + p\,dV + \frac{V}{\gamma}dp = c_V dT \quad \text{or} \quad V\,dp = c_p dT \tag{2.12}$$

Substituting for V from (2.7) and re-arranging to get[c]

$$\frac{dT}{dp} = \frac{R_d T}{c_p p}. \tag{2.13}$$

---

[b]See Sec. 6.2.1 for a more detailed discussion or refer to a standard textbook on thermodynamics, e.g. Ref. 2.

[c]Note for unit mass, we can write $V = 1/\rho$.

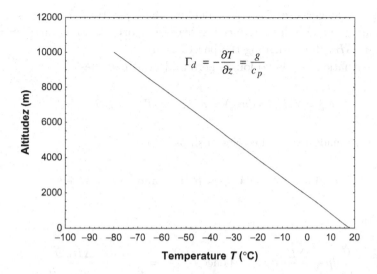

Fig. 2.3   Adiabat for dry air calculated assuming a sea level temperature of 18°C.

We finally obtain the lapse rate from (2.8) and (2.13) using the chain rule

$$\Gamma_d = -\frac{\partial T}{\partial z} = -\frac{\partial T}{\partial p} \times \frac{\partial p}{\partial z} = \frac{R_d T}{c_p p} \times \frac{gp}{R_d T}$$

$$\Gamma_d = -\frac{\partial T}{\partial z} = \frac{g}{c_p}.$$

(2.14)

This is the *dry adiabatic lapse rate* (Γ_d) and implies that for dry air, the temperature decreases linearly with altitude (Fig. 2.3). The specific heat of dry air[3] is $1008\,\mathrm{Jkg^{-1}\,K^{-1}}$ and therefore $\Gamma_d = 9.8 \times 10^{-3}\,\mathrm{Km^{-1}}$ or about 1°C for each 100 m of ascent. The graph of altitude versus temperature is known as an adiabat.

2.3.2 *Saturated lapse rate and vertical stability*

In practice air invariably contains some water vapour (only in the most extreme environments — hot deserts, polar ice caps — is it effectively dry) and the dry adiabatic lapse rate only represents the limiting case. However, provided that the air is not saturated, then (2.14) may be used with a corrected value of $c_p = c_{Pef}$. If the mass fraction of water vapour is w_m, then

$$c_{p_{ef}} = w_m c_{p_w} + (1 - w_m)c_{p_d}$$

(2.15)

where c_{p_w} and c_{p_d} are the specific heats at constant pressure for water vapour and dry air respectively.

A rising parcel of moist air will cool, the relative humidity will increase and at some altitude the air will become saturated. Some of the moisture will condense and we can no longer use (2.15) to calculate the lapse rate. The mass fraction of water vapour $(-dw_m)$ will be reduced, releasing a corresponding amount of heat of evaporation, ΔH_v thus warming the parcel of air.

In this situation, δQ is no longer zero and (2.9) becomes

$$\delta Q = \Delta H_v(-dw_m) = c_v dT + pdV = c_p dT - Vdp \qquad (2.16)$$

where we have made use of the relationships

$$d(pV) = pdV + Vdp = RdT \quad \text{and} \quad c_p = R + c_v. \qquad (2.17)$$

From (2.16),

$$dT = \frac{V}{c_p}dp - \frac{\Delta H_v}{c_p}dw_m \quad \text{hence} \quad \frac{\partial T}{\partial z} = \frac{V}{c_p}\frac{\partial p}{\partial z} - \frac{\Delta H_v}{c_p}\frac{\partial w_m}{\partial z}. \qquad (2.18)$$

Replacing V by $(1/\rho)$ and taking $(\partial p/\partial z)$ from (2.4) gives for the saturated adiabatic lapse rate

$$\Gamma_{sat} = -\frac{\partial T}{\partial z} = \frac{g}{c_p} + \frac{\Delta H_v}{c_p} \times \frac{\partial w_m}{\partial z} \approx \Gamma_d + \frac{\Delta H_v}{c_p} \times \frac{\partial w_m}{\partial z}. \qquad (2.19)$$

The saturated lapse rate is lower than the dry case[d] and the air cools more slowly as it rises, due to the release of heat of evaporation. In addition, Γ_{sat} will not be linear as the rate at which water vapour condenses i.e. the last term in (2.19) will not be a constant.

The rate at which the parcel of air cools (i.e. lapse rate) as it ascends is important in determining the vertical stability of the atmosphere. Suppose, for example, that the parcel is displaced with respect to the surrounding air (assumed to be in hydrostatic equilibrium) and on displacement the net resulting force acts to restore it to its initial position, then the parcel is in equilibrium with the environment. The atmosphere is said to be stable. If on the other hand, the net force acts to accelerate the parcel away from its initial position, then the system is in unstable equilibrium.

In the latter case, the parcel is cooling more slowly than the environment giving rise to positive buoyancy forces. The air is rising, resulting in cloud formation and possibly rain. This is termed *convective instability*. Conversely, when the parcel is in equilibrium with the atmosphere, it is cooling more rapidly and experiencing negative buoyancy forces.

---

[d]As water condenses from the vapour, the water vapour fraction is reduced making $(\partial w_m/\partial z)$ negative.

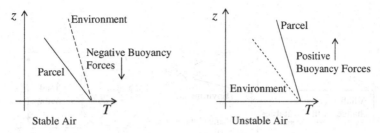

Fig. 2.4 Adiabats for stable and unstable atmospheres; in both instances the environment could be saturated.

We represent these situations by comparing the lapse rates with that of the environment, Γ_{env}. The situation may be summarised by the following criteria

$$\Gamma_{env} > \Gamma_{parcel} \quad \text{unstable atmosphere;}$$
$$\Gamma_{env} = \Gamma_{parcel} \quad \text{neutral atmosphere;}$$
$$\Gamma_{env} < \Gamma_{parcel} \quad \text{stable atmosphere.}$$

These conditions apply irrespectively of whether Γ_{env} is the saturated lapse rate for the environment. The case for saturated air is illustrated in Fig. 2.4, which demonstrates that even when the environment is saturated, the atmosphere may still be stable, provided $\Gamma_{parcel} > \Gamma_{env}$.

The temperature gradient of the surrounding environment through which the parcel is moving is a complicated function of vertical buoyancy forces, local geography, and also the horizontal motion of the air (see Sec. 2.4). Consequently, the dry (and saturated) adiabats discussed here are considerable simplifications of reality.

2.3.3 *Cloud formation and precipitation*

Broadly speaking, in an unstable atmosphere, water vapour will start to condense out of rising air forming clouds when the local relative humidity reaches 100%. The altitude at which this occurs is referred to as the *convection condensation level* (CCL) or the *lifting condensation level* (LCL) and corresponds to the cloud base. Condensation in a parcel of air at the CCL will release latent heat to make the parcel warmer than its surroundings and it will therefore follow the saturated adiabat (Fig. 2.5) as it continues to ascend and expand adiabatically.

There are three sources of energy for the work of expansion:

  i)   The internal energy of the parcel of air
  ii) The latent heat of condensation
 iii) The cooling liquid

In saturated air, the energy released during the change of phase provides much of the work of expansion. In contrast, all the work of expansion for dry or unsaturated

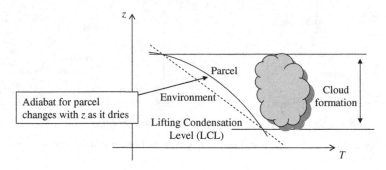

Fig. 2.5   Cloud formation as a parcel of air ascends through the LCL.

air must be supplied from the internal energy of the gas, and thus a parcel of dry air will cool more rapidly than one that is saturated.

The process of cooling and condensation of saturated air is only adiabatic for the entire system, i.e. air, water vapour and condensates. It is not adiabatic for the individual constituents, nor will it be adiabatic if any of the condensates fall out of the cloud as rain or snow, since the process then becomes irreversible with energy being removed by the precipitation. However, for present purposes, we shall assume the expansion to be both reversible and adiabatic.

In condensation, water droplets nucleate on aerosol particles in the atmosphere, typically, sea salt spray, smoke, dust etc. A volume of air that contains no aerosols at all, can in principle sustain levels of supersaturation of several hundred percent and the nucleating particles are therefore a critical element in the process. They range in size from a few nm to a few μm and are usually present in densities of $10^9 - 10^{10}\,\mathrm{m}^{-3}$. As a result, a typical cumulus cloud will contain a comparable density of water droplets $(10^8\,\mathrm{m}^{-3})$ with a mode radius of a few μm. This phase of the adiabatic expansion and cloud development is sometimes referred to as the rain stage, although no rain may have actually fallen.

The water droplets accelerate downwards until they reach their terminal velocity, when the downwards gravitational force just balances the opposing viscous (frictional) force, which gradually increases as the droplets encounter air of greater density. Once the droplets fall beneath the CCL, i.e. beneath the cloud, they start to re-evaporate and the vapour will return to the cloud. Precipitation during the rain stage will only take place if the droplet size becomes large enough to reach the ground without re-evaporation. In practice, this requires a droplet radius of $\sim 1\,\mathrm{mm}$. Water droplets can grow to this size by coalescence with other droplets.

If there are sufficient up draughts within the cloud, then small droplets will be carried upwards. Expansion and cooling continue until the temperature drops to the freezing point and all the liquid water freezes. The process is now isothermal, the temperature remaining at freezing, and the only available energy for the work of expansion is the latent heat of fusion. This is the hail stage.

Continued ascent and expansion results in the snow stage, where water vapour sublimates directly to ice. Thermodynamically, this is the same as for the rain stage. The energy for the work of expansion is being provided by the change in the internal energy of the air, the latent heat of sublimation and the cooling ice.

At cloud temperatures below about $-30°C$, water vapour may condense to form supercooled liquid and both ice crystals and water droplets co-exist in the same cloud. Water vapour surrounding the ice crystals will be supersaturated with respect to the ice crystal and will freeze directly onto them. The ice crystals grow rapidly in this way and then start to fall through the cloud colliding with small water droplets along the way. These freeze immediately on impact and the crystals start to grow by accretion. Aggregates of such crystals form by coalescence with other crystals to produce snow flakes. When they fall below the cloud, the flakes will melt and fall as rain unless the temperature remains below freezing point. Most of the rain in the middle and higher latitudes starts off as snow.

If the up draughts in low level (cumulonimbus) clouds are particularly strong, then ice particles formed in the hail stage may be repeatedly moved up and down within the cloud with alternating conditions of freezing and supercooling. The ice particle grows with concentric layers of ice and frost until it is massive enough to fall as hail stones.

The higher cirrus clouds are much colder $(-40°C)$ and are composed entirely of small ice crystals rather than water droplets. Ice crystals falling from the high altitude cirrus clouds evaporate more slowly than do water droplets falling from low altitude cumulus clouds, which as a result, have clear distinct edges. By comparison, cirrus clouds are diffuse and indistinct.

2.4 Horizontal Motion of the Air

2.4.1 *General circulation*

The horizontal motion of air is generally what we perceive as 'wind'. The ambient temperature and humidity etc. is determined in large part by the horizontal motion of the air, bringing in warmer/cooler or dryer/wetter air as the case may be. In deriving an understanding of the behaviour of winds and the driving forces, we shall assume independent horizontal frictionless flow. This is really only valid at heights above the ground of about 500 m where wind speeds are largely unaffected by the terrain.

The prevailing winds at some geographical location are a consequence of the large scale circulation of the atmosphere. In order to obtain an appreciation of the larger scale dynamics, we shall assume that the Earth is longitudinally uniform. Evidently, this is not the case, but the concept is helpful in providing a description of the various high and low pressure regions (Fig. 2.6) and the corresponding wind patterns. Intense solar heating of the surface of the Earth at the equator warms the air which rises, drawing in replacement air from the higher latitudes. The rising equatorial air cools, moves polewards and starts to descend, creating

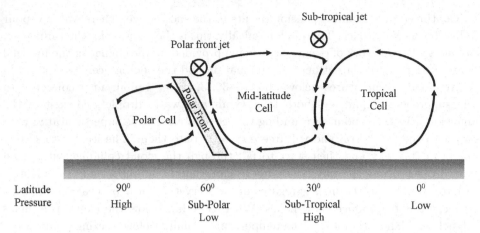

Fig. 2.6   Three cell meridional general circulation model.

the subtropical high pressure region. The descending air mass spreads out towards both the equatorial and polar regions. Air moving towards the pole will eventually encounter the much colder polar air mass creating the polar front, where it rises creating the subpolar low pressure belt. The polar air mass also rises and returns to the pole. The mean atmospheric pressure is high at the pole and low at the equator. The result is to form three meridional cells of circulation illustrated in Fig. 2.6: the tropical or Hadley cell,[e] the mid-latitude cell and the polar cell.

In essence, the prevailing wind patterns are determined by the interaction of the pressure belts with the rotation of the Earth (see Sec. 2.4.2). In a longitudinally uniform *non-rotating* Earth, the prevailing winds would blow in north-south directions in response to the meridional circulation. The rotation of the Earth acts to deflect the winds so that they blow in more easterly and westerly directions. In the northern hemisphere, the prevailing winds between the equator and the subtropical high pressure belt are the easterly Trade Winds (so called because they enabled trade to take place between the 'old' and 'new' worlds in the days of sail). Between the subtropical high and subpolar low pressure belts, the prevailing winds are the zonal westerlies, bringing wet and warm weather to Western Europe for example. North of the sub-polar low pressure belt, the prevailing winds are cold polar easterlies. A similar situation exists in the southern hemisphere.

2.4.2 *Forces driving horizontal motion*

To determine the equation motion of the air constituting the wind, we consider a volume element of air, dV with mass ρdV and velocity $\underline{u}$ and sum the forces acting

---

[e] After Hadley who first observed it in 1735.

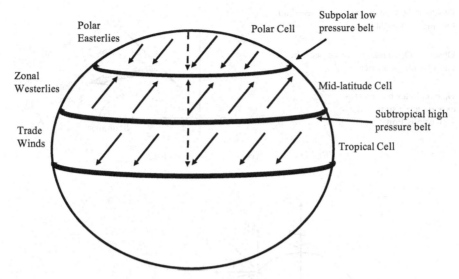

Fig. 2.7 Prevailing wind patterns in a three cell meridianal general circulation model for a longitudinally uniform rotating Earth. (Dashed arrows indicate wind directions in a non-rotating Earth.)

on the element in the usual way.

$$\frac{du}{dt}\rho dV = \underline{F}_p + \underline{F}_C \qquad (2.20)$$

where $\underline{F}_p$ and $\underline{F}_C$ are the pressure gradient and Coriolis forces respectively. In principle, there will also be viscous drag forces which arise when neighbouring layers of air are moving at different velocities. At heights of more than $\sim 500\,\mathrm{m}$ above the ground, the vertical gradient of wind velocity is small and hence the drag is also small and may be neglected.

The Coriolis force arises from the rotation of the Earth. Newton's laws only apply in inertial reference frames i.e. reference frames that are not under acceleration. To make it possible to use Newton's Second Law in an accelerating or non-inertial reference frame, it is necessary to introduce some fictitious force (pseudo force) to account for the acceleration of the reference frame. The Coriolis force is one such pseudo force that arises when the reference frame is rotating. The situation is illustrated in Fig. 2.8. Observer $\mathbf{O_S}$ in an inertial frame sees a particle moving in a straight line with a constant velocity $\underline{u}$ and infers the particle is not experiencing any force. On the other hand, observer $\mathbf{O_R}$ moving in a rotating frame sees the same particle moving in a curved trajectory and concludes therefore that it must be under the action of some force.

A coordinate system fixed to the Earth's surface is rotating with the Earth and is therefore not an inertial system. A parcel of air moving due north would appear to an observer on the ground to be moving along an arc towards the west.

Observer O_S in stationary frame sees ball moving in straight line:

Observer O_R in rotating frame sees the ball moving in an arc:

O_R concludes ball must be experiencing some force — Coriolis Force.

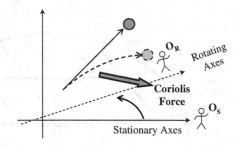

Fig. 2.8　Illustration of Coriolis force.

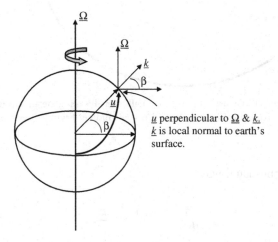

u perpendicular to $\underline{\Omega}$ & $\underline{k}$. $\underline{k}$ is local normal to earth's surface.

Fig. 2.9　Geometry for calculating the tangential component of the Coriolis force.

The Coriolis force, $\underline{F}_C$ acting on a mass m moving with linear (tangential) velocity $\underline{u}$ is given by[4]

$$\underline{F}_C = -2m(\underline{\Omega} \times \underline{u}) \qquad (2.21)$$

where $\underline{\Omega}$ is the angular velocity of the rotating reference frame. The cross product indicates that $\underline{F}_C$ will be directed along the normal to the plane containing $\underline{u}$ and $\underline{\Omega}$.

We are interested in the perceived horizontal (i.e. directions tangential to the surface of the Earth) motion of the wind and must therefore use the tangential component of the Coriolis force for which we need to use the local vertical $\underline{k}$ (termed the zenith) component of $\underline{\Omega}$. Except at the poles, the direction of the $\underline{k}$ will not be the same as $\underline{\Omega}$ (which is directed towards the north along the axis of the Earth's rotation) and thus in general the vertical component of the angular velocity component depends on the latitude (β) and is given as (Fig. 2.9)

$$\underline{\Omega}_{\text{Vertical}} = \sin(\beta)\Omega\underline{k} \qquad (2.22)$$

The derivation of the pressure gradient force is illustrated in Fig. 2.10.

$$dV = dxdydz$$

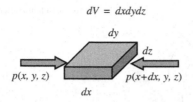

Fig. 2.10   Derivation of the pressure gradient force.

The pressure gradient force along the x-direction is the difference in the pressures on the left and right sides of the elemental volume.

$$F_p = p(x.y.z)dydz - p(x+dx, y, z,)dydz$$

$$= p(x, y, z)dydz - \left\{ p(x, y, z) + \frac{\partial p}{\partial x}dx \right\} dydz$$

$$F_p = -\frac{\partial p}{\partial x}dV \quad \text{or} \quad \underline{F}_p = -\nabla p dV \quad \text{in three dimensions.} \qquad (2.23)$$

Substituting for the forces in (2.20)

$$\frac{d\underline{u}}{dt}\rho dV = -\nabla p dV - 2\sin(\beta)\Omega(\underline{k} \times \underline{u})\rho dV \qquad (2.24)$$

gives us the equation of motion for the elemental volume.

2.4.3 *Geostrophic flow*

Equation (2.24) is the equation of horizontal motion for frictionless flow under the assumption of hydrostatic equilibrium. It is worth noting in passing that the neglect of friction (and gravity) forces implies the absence of tangential and centripetal components of acceleration. Under these circumstances, flow is said to be *geostrophic*. At altitudes above ~ 500 m, not only is viscous drag small (as assumed in Sec. 2.4.2) but under ordinary conditions, the wind velocity varies slowly over time ($d\underline{u}/dt \approx 0$). The left hand side of equation (2.24) is therefore effectively zero. Consequently,

$$2\rho\sin(\beta)\Omega(\underline{k} \times \underline{u}_g) = -\nabla p \qquad (2.25)$$

where $\underline{u}_g$ is the *geostrophic wind velocity*.

Equation (2.25) represents a balance between the Coriolis and pressure gradient forces. The cross product on the left is a vector that must obviously have the same direction as the negative of the pressure gradient (i.e. from high to low pressure) and hence $\underline{u}_g$ and $\underline{k}$ must both be perpendicular to it. Since by definition $\underline{k}$ is in the vertical or zenith direction, the wind direction must be parallel to the isobars. In the northern hemisphere, $\underline{k}$ is directed upwards and wind will blow with high pressure to the right (viewed from above) explaining, for example, the prevailing

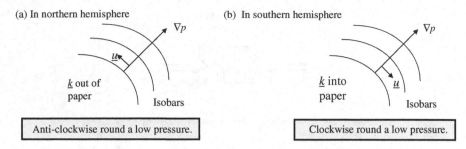

Fig. 2.11 Geostrophic flow around low pressure systems in (a) northern hemisphere and (b) southern hemisphere.

wind directions either side of the subtropical high pressure belt. It should be noted that this is a somewhat simplified view of the situation and local high and low pressure systems often have a more immediate impact on weather.

In the northern hemisphere, winds blow anticlockwise around a low pressure system (Fig. 2.11) and clockwise around a centre of high pressure, as will be apparent on television weather forecasts. In the southern hemisphere, the local zenith points southward, the direction of $\underline{k}$ is reversed and geostrophic winds blow in the opposite sense.

The geostrophic wind speed is directly proportional to the pressure gradient and varies with latitude

$$u_{\mathrm{g}} = \frac{|\nabla p|}{2\rho \sin(\beta)\Omega} = \frac{|\nabla p|}{f\rho} \tag{2.26}$$

where f is the Coriolis parameter. Not surprisingly, wind speed increases in direct proportion to the pressure gradient. It also increases with increasing latitude.

2.4.4 *Vertical wind shear*

Assuming that the angular velocity of air at different altitudes remains constant, then the tangential (i.e. horizontal) wind speed must increase proportionately. The difference in wind speed between neighbouring layers of the atmosphere is known as *wind shear* and we define the *shear vector* to represent the vector difference in the wind velocity between adjacent layers. The shear vector acts to increase the westerly component of the wind throughout the troposphere. The shear vector in a given layer is controlled by the horizontal temperature gradient and is often called the thermal wind as a result.

To derive the vertical distribution of geostrophic wind velocity, we substitute for the pressure in equation (2.25) from the equation of state (2.5), and noting that for unit mass, $(1/V) = \rho$, the density:

$$\underline{k} \times \underline{u}_g = -\frac{1}{f\rho}\nabla p = -\frac{R_{air}}{f\rho}\{T\nabla\rho + \rho\nabla T\} \tag{2.27}$$

where R_{air} is the specific gas constant for air. Dividing throughout by T and using the identity

$$\frac{1}{y}\frac{\partial y}{\partial x} = \frac{\partial}{\partial x}(\ln y)$$

$$\frac{k \times u_g}{T} = -\frac{R_{air}}{f}\left\{\frac{\nabla\rho}{\rho} + \frac{\nabla T}{T}\right\}$$

$$= -\frac{R_{air}}{f}\{\nabla\ln\rho + \nabla\ln T\}$$

$$= -\frac{R_{air}}{f}\nabla\ln p. \tag{2.28^f}$$

To find the variation of wind velocity with altitude, we differentiate equation (2.28) with respect to z

$$\frac{\partial}{\partial z}\left\{\frac{k \times u_g}{T}\right\} = -\frac{R_{air}}{f}\frac{\partial}{\partial z}\nabla\ln p = -\frac{R_{air}}{f}\nabla\frac{\partial}{\partial z}\ln p. \tag{2.29}$$

To evaluate the last term, we follow a similar procedure using the hydrostatic equation (2.4) as a starting point

$$\frac{1}{\rho}\frac{\partial p}{\partial z} = \frac{R_{air}}{\rho}\frac{\partial}{\partial z}(\rho T) = \frac{R_{air}}{\rho}\left\{\rho\frac{\partial T}{\partial Z} + T\frac{\partial\rho}{\partial z}\right\} = -g \tag{2.30}$$

dividing throughout by T etc.

$$-\frac{g}{T} = R_{air}\left\{\frac{1}{T}\frac{\partial T}{\partial z} + \frac{1}{\rho}\frac{\partial\rho}{\partial z}\right\} = R_{air}\left\{\frac{\partial}{\partial z}\ln T + \frac{\partial}{\partial z}\ln\rho\right\} = R_{air}\frac{\partial}{\partial z}\ln p \tag{2.31}$$

substituting in equation (2.29)

$$\frac{\partial}{\partial z}\left\{\frac{k \times u_g}{T}\right\} = \frac{g}{f}\nabla\frac{1}{T} = -\frac{g}{fT^2}\nabla T. \tag{2.32}$$

The cross product on the left only has horizontal components because k is in the vertical and we may therefore conclude from equation (2.32) that the variation in geostrophic wind velocity depends on the horizontal temperature gradient. Implicit in the above analysis is the fact that the surfaces of constant pressure (isobaric) do not coincide with the surfaces of constant density. This model of the atmosphere is termed *baroclinic*.

From equation (2.8) it is clear the change of pressure with altitude is mainly dependent on the temperature. On average, we would therefore expect isobaric

---

[f]Note: Since R_{air} is a constant, its derivative is zero. i.e. $\nabla\ln R_{air} = 0$ and we can add it to the third term in equation (2.28). Then, we express the equation of state as $\ln p = \ln\rho + \ln T + \ln R_{air}$: $\nabla\ln p = \nabla\ln\rho + \nabla\ln T + \nabla\ln R_{air}$.

surfaces to be tilted downwards towards the poles since these are cooler. The horizontal pressure gradient therefore changes in magnitude and the direction changes with height as a result. Wind speeds continue to increase with altitude, reaching their maximum near the top of the troposphere. Above the tropopause, the temperature gradient reverses sign and thermal winds start to decrease.

The vertical distribution of wind shear varies with latitude and season. The temperature distributions in the upper troposphere favour the formation of belts of very strong winds called jet streams. These are steady high speed winds that blow in a westerly direction around the globe. There are two main jets, the subtropical jet and the polar front jet (Fig. 2.6). The northern hemisphere subtropical jet is located at an altitude of about 200 mbar (~ 13 km) and in summer is located at a latitude of 40° N and blows at a mean speed of about $20\,\mathrm{ms}^{-1}$ (70 kph). It moves to lower latitudes in the winter months. The polar front jet is strongly correlated with the location of the polar front and is located essentially over the sea level position of the polar front at an altitude of ~ 500 mbar (11 km). Wind speeds in the polar front jet in mid-winter are much stronger than those of summer $\sim 50\,\mathrm{ms}^{-1}$ (180 kph).

2.4.5 *Horizontal wind shear — weather fronts*

It will be evident from the admittedly simplified picture of general circulation, Fig. 2.6, that there appears to be quite distinct circulations of air which remain separate. To a surprising degree, different bodies of air retain physical properties (temperature, humidity) that are uniform across a given pressure (altitude) level. Such bodies of air are referred to as *air masses* and acquire their particular properties in source regions characterised by uniformity of temperature and pressure, for example, in polar regions, tropical seas, deserts, continental plains etc.

Meteorological observations provide strong evidence that the transition from one air mass to another is essentially discontinuous. In reality, there is a *zone of transition*, but this is a relatively small distance (~ 100 km) compared to the scale of atmospheric motion, ~ 1000 km.

The boundary between two air masses is known as a *front* and the transition region as the *frontal zone*, which for convenience may be treated mathematically as a surface of discontinuity. The colder (denser) zone generally forms a wedge under the warmer (less dense) zone so that the frontal zone is inclined to the surface of the Earth (Fig. 2.12) and much of the weather in the mid-latitudes is determined by the relative motions of warm and cold air masses. As the warmer subtropical air mass rides up over the colder polar air, it cools, and the humidity increases, eventually leading to steady light rain and drizzle.

Instabilities on the polar front, separating the cold polar air mass and the warm subtropical air mass, lead to the development of mid-latitude depressions.

The life cycle of a typical depression lasts a few days during which it may travel several thousand kilometres. The cycle is illustrated in Fig. 2.13. Prior to the formation of the depression, the polar front lies roughly in an east-west

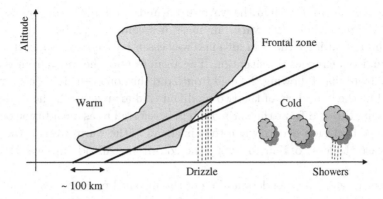

Fig. 2.12   Vertical section through an idealised frontal zone between cold and warm air masses.

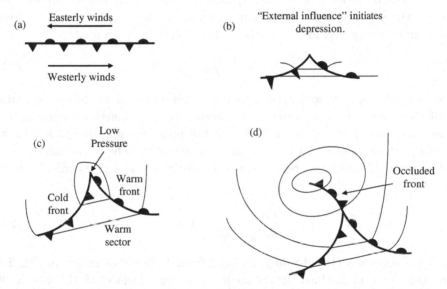

Fig. 2.13   Life cycle of a typical mid-latitude depression in the northern hemisphere: (a) quasi-stationary section of polar front (arrows indicate wind directions); (b) external influence initiates a low pressure region (lines represent isobars); (c) the low pressure develops into a full depression; (d) the cold front 'overtakes' the warm front producing an occluded front.

direction, with (in the northern hemisphere) easterly winds to the north and westerly winds to the south maintaining the front more or less in equilibrium. An external perturbation at some point will create a local instability called a wave, which will start to move in a westerly direction (in the northern hemisphere) becoming more pronounced as it does so generating a low pressure region at its core — the depression. The leading edge of the wave, where cold polar air is being displaced by warmer subtropical air, is termed the *warm front*, while the trailing edge (warm air replaced by cold air) the *cold front*. The region enclosed by the warm and cold fronts is composed of warm air and is therefore called the *warm sector*. The cold front travels more rapidly than the warm front and gradually overtakes it, forming an *occluded front*, where the warm air mass lies completely above the colder air mass. Although initially the cold air mass was essentially homogenous, the passage of the warm front changes the situation. The air in front of the warm front is affected differently from that following the cold front, and thus on occlusion a new front will be created at the boundary of the two modified cold air masses. If the coldest air is that following the cold front the entire system is referred to as a cold-type occlusion. A warm-type occlusion occurs when the air ahead of the warm front is the coldest, in which case the less cold air following the cold front will ride up over the coldest air.

In order to gain some understanding of the physical processes, we shall assume that the front may be treated as a surface of discontinuity (as suggested above). In particular, we define an ideal front as one described by a discontinuity in the density, i.e. where the densities of the warm (ρ_w) and cold (ρ_c) air masses are different. However, the pressure must be continuous across the front, otherwise there would be an infinite pressure gradient. Mathematically, this is represented as

$$\rho_w - \rho_c \neq 0; \quad p_w - p_c = 0 \qquad (2.33)$$

where the subscripts w and c refer to the warm and cold air masses on either side of the frontal zone respectively. Equation (2.33) is sometimes referred to as the *dynamic boundary condition*. Although the pressure must be continuous, the pressure gradient is not so constrained, and may be taken to be discontinuous across an ideal front. Defining the x-direction to be horizontal and perpendicular to the front, then

$$\left.\frac{dp}{dx}\right|_w - \left.\frac{dp}{dx}\right|_c \neq 0. \qquad (2.34)$$

Consider a vertical section through an ideal frontal system as shown in Fig. 2.14, where the front is inclined at an angle ϕ_f to the surface of the Earth. We construct an infinitesimal rectangle ABCD on the frontal surface as shown, such that

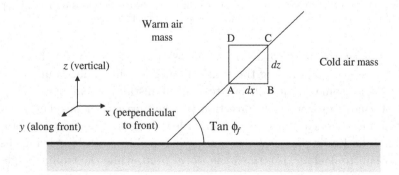

Fig. 2.14 Vertical section through an idealised front for the derivation of Margule's formula.

AB = CD = dx and AD = BC = dz. The slope is just $dz/dx = \tan(\phi_f)$. The pressure gradients in the warm and cold sectors close to the front may be written as

$$\left.\frac{dp}{dx}\right|_w = \frac{p_C - p_D}{dx} \quad \text{and} \quad \left.\frac{dp}{dx}\right|_c = \frac{p_B - p_A}{dx}. \tag{2.35}$$

To first order,

$$p_D = p_A + \frac{\partial p}{\partial z}dz; \quad \left.\frac{dp}{dx}\right|_w = \frac{p_C - p_A - (\partial p/\partial z)\,dz}{dx} = \frac{p_C - p_A + g\rho_w dz}{dx} \tag{2.36}$$

where we have replaced the term $(\partial p/\partial z)$ using the hydrostatic equation (2.4). Following a similar approach for the pressure gradient function on the cold side, we get for the difference in pressure gradients

$$\left.\frac{dp}{dx}\right|_c - \left.\frac{dp}{dx}\right|_w = g\,(\rho_c - \rho_w)\frac{dz}{dx} \quad \text{or} \quad \frac{dz}{dx} = \frac{1}{g\,(\rho_c - \rho_w)}\left\{ \left.\frac{dp}{dx}\right|_c - \left.\frac{dp}{dx}\right|_w \right\}. \tag{2.37}$$

Assuming that the winds are geostrophic with speeds given by equation (2.26), then in 1-dimension

$$\frac{dp}{dx} = u_g f\rho \quad \text{and hence} \quad \frac{dz}{dx} = \frac{f}{g}\left\{ \frac{u_{g_c}\rho_c - u_{g_w}\rho_w}{\rho_c - \rho_w} \right\}. \tag{2.38}$$

Finally, we substitute for the densities using the ideal gas law (2.5) expressed in terms of per unit mass, where the volume is equivalent to the reciprocal of the density ($V \equiv 1/\rho = p/R_{air}T$).

$$\frac{dz}{dx} = \frac{f}{g}\left\{ \frac{u_{g_c}T_w - u_{g_w}T_c}{T_w - T_c} \right\} = \frac{f\overline{T}}{g}\left\{ \frac{u_{g_c}(T_w/\overline{T}) - u_{g_w}(T_c/\overline{T})}{T_w - T_c} \right\} \tag{2.39}$$

where we have made use of the dynamic boundary condition (2.33) and have assumed that the specific gas constant R_{air} is the same in the two air masses, and $\overline{T}$ is the mean temperature. Since the temperatures in the warm and cold sectors will be similar and much greater than $0\,\mathrm{K}$ then $T_c/\overline{T} \approx T_w/\overline{T} \approx 1$, and

$$\tan(\phi_f) = \frac{dz}{dx} \approx \frac{f\overline{T}}{g}\left\{ \frac{u_{g_c} - u_{g_w}}{T_w - T_c} \right\}. \tag{2.40}$$

Equation (2.40) is the Margules' formula for the slope of the front and shows that the slope is directly proportional to the wind shear and inversely proportional to the temperature difference. In practice, it is found that equation (2.40) underestimates $\tan(\phi_f)$ for cold fronts and overestimates it for warm fronts. Additionally, for a given type of front, $\tan(\phi_f)$ is observed to vary little irrespective of the temperature contrast across the front. This implies that the wind shear and temperature contrast are proportional to each other, i.e. a larger temperature contrast gives rise to a greater wind shear.

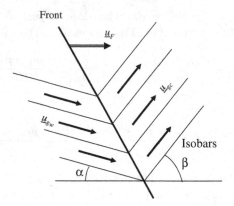

Fig. 2.15   Distribution of isobars at a front, showing the frontal velocity $\underline{u}_f$ (arrows indicate geostrophic winds).

The energy for the formation and movement of a frontal depression is ultimately derived from the sun. Solar energy is converted into atmospheric heat energy, either directly by absorption or more usually through the processes of evaporation and condensation that absorb and then release latent heat. Across a front where warm air is adjacent to colder air, the distribution is uneven and represents a concentration of potential energy. As the depression develops, some of the available potential energy is converted to kinetic energy carried by the increased wind speeds, an exchange implied in the Margules' formula (2.40) which associates the wind speeds with the temperature difference. Once the front has become fully occluded, the source of potential energy is no longer available, and frictional losses eventually bring the circulation to a halt.

Although they may be stationary, in general frontal systems move. To a first approximation, the velocity will be equal to the horizontal component of either of the geostrophic wind components. Figure 2.15 shows the distribution of isobars at an imaginary front, with the warm sector in this case on the left and the low pressure centre towards the top of the diagram.[g] The isobars meet at the front as required (2.33) but change direction abruptly at the front as a result of the pressure gradient discontinuity (2.34). The geostrophic winds run parallel to the isobars (Sec. 2.4.3). Continuity requires that the front must move in a direction such that

$$\underline{u}_F = \underline{u}_{g_w} \cos(\alpha) = \underline{u}_{g_c} \cos(\beta) \tag{2.41}$$

where α and β are the angles between the direction of frontal motion and the isobars on the warm and cold sides respectively. We can see from equation (2.41) why there is often an abrupt change in the wind direction as a front passes, particularly the cold front as this tends to bring cold polar air circulating round the low pressure centre. The associated weather is referred to as cyclonic as a result.

---

[g]Viewed in plane, i.e. as it would be on a weather map.

2.4.6 *Tropical cyclones — hurricanes*

In equatorial zones, the winds from the two hemispheres converge in a band of rising air known as the Intertropical Convergence Zone (ITCZ). Over tropical oceans, where the air is moist, the rising air will be unstable causing the formation of cumulonimbus clouds and consequently strong convection. The resulting local convergence of winds will rise and spiral up under the action of the Coriolis force[h] to form the nucleus of a major storm.

These drift westwards and may intensify into fully mature *hurricane*[i] storms of great spatial extent. Although the converging air spirals inwards and upwards with increasing speed towards the storm centre, most of the incoming air never reaches the core which remains calm and rain-free by comparison. This is the well known 'eye' of the hurricane. Around the eye, wind speeds are extremely high (typically in excess of 110 kph) and destructive. The incoming air is made very moist by rapid forced evaporation from the tropical oceans and as it rises forms extensive spirals of cumulonimbus clouds, with accompanying torrential rain and thunder storms (Fig. 2.16). On reaching the tropopause, the air diverges away from the storm centre.

Tropical hurricanes generally move slowly in a westerly direction and slightly polewards (i.e. away from the equator), although not necessarily in a smooth trajectory, making prediction of their future direction difficult. They are storms of immense size extending over many hundreds of square kilometres. The eye may be 40 km across.

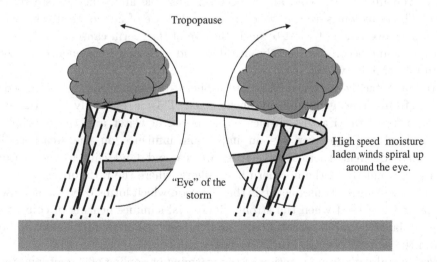

Fig. 2.16   Schematic diagram of a section through a hurricane. Moisture laden air is drawn in towards the centre, where it spirals upwards generating destructively high winds and torrential rain.

---

[h]Tropical cyclones do not form at latitudes less than ∼ 5°, because here the Coriolis force is negligible.
[i]In the western Atlantic. In the western Pacific they are known as *typhoons*.

Hurricanes are driven by the latent heat released on condensation and which is ultimately derived from the warm oceans. Consequently, they are most likely to occur at the end of summer when the seas are at their warmest. On landfall (typically on the eastern seaboard of the USA), hurricanes quickly decline and die away as they move inland and the energy source is removed. This is in contrast to cyclonic weather at higher latitudes (particularly in the northern hemisphere) where the kinetic energy is derived from the temperature contrast across weather fronts (Margules' formula (2.40)). However, in both instances the fundamental energy source derives from the Sun.

2.5 Summary

This chapter has concentrated on some of the ways in which the vertical and horizontal motions of air in the atmosphere determine the weather. In it, we have attempted to show how the rate at which rising air cools (lapse rate) determines whether the atmosphere is likely to be stable or alternatively unstable, in which case clouds will form and rain may follow. Horizontal motion of the atmosphere is most clearly experienced as wind, the velocity of which is the consequence of the interaction of pressure gradients and the rotation of the Earth as expressed in the Coriolis force. Frontal weather systems may be seen as the coming together of both the horizontal and the vertical behaviour of the atmosphere. As the warm air is driven up over the cold air, it becomes unstable and generally steady rain follows. The situation is even more extreme in the case of hurricane storms, where horizontal convergence of wind towards the eye of the storm causes the air to rise very rapidly and become extremely unstable, and with as a consequence torrential rain and destructive winds.

We have tacitly assumed that atmospheric dynamics can be discussed in isolation from other geographical phenomena. This is evidently not the case. The majority of the Earth's surface is composed of water and the distribution between land and sea exercises an important influence on the weather. The significantly different thermal properties of sea and land cause large seasonal changes, particularly in the northern hemisphere, where the ratio of sea to land is about 60:40.[j] A direct consequence is the monsoon which forms over the northwest of India and is caused when the strong differential summer heating of continental Asia is replaced by rapid radiative cooling. The reversal of air flow draws in moist air from the warmer oceans, leading to heavy rainfall.

We have also made no mention of the warming or cooling effects of long range circulatory ocean currents, or of periodic and intermittent events such as the El Niño. Interactions between ocean and atmosphere control weather on a global scale and we shall discuss these in the next chapter on global climate.

---

[j]In the southern hemisphere the sea:land ratio is nearer 80:20.

2.6 Problems

1. Assuming the air is dry, estimate the temperature and pressure (in Pa) at the top of the troposphere, if at sea level the temperature is 293 K and the pressure is 1 atmosphere.

2. Water vapour makes up 1% by weight of a sample of moist air. What is the mixing ratio? What is the unsaturated adiabatic lapse rate for this parcel of air? (The specific heat of water vapour at constant pressure is $1810\,\mathrm{JKg^{-1}K^{-1}}$.)

3. The six o'clock shipping forecast reports that a low pressure of 980 mbar is centred out in the Atlantic at a position (latitude 54° 30'north; longitude 16° west). At the same time a weather ship in the North Sea (latitude 54° 30' north; longitude 2°15' east) records a geostrophic wind speed of $12.2\,\mathrm{ms^{-1}}$ from due south. What is the local pressure at the weather ship? (radius of the Earth = 6370 km, mean density of air = $1.2\,\mathrm{kg\,m^{-3}}$).

4. A warm front moving due west along a line of latitude of 55°, makes an angle of 0.37° with the surface of the Earth. If the temperatures in the warm and cold sectors are 8°C and 3°C respectively, determine the difference in the geostrophic wind speeds. If the isobars in the cold sector make an angle with the front twice that of the isobars in the warm sector and the geostrophic wind in the warm sector is $8\,\mathrm{ms^{-1}}$, how rapidly is the front moving?

References

1. R. C. West, *Handbook of Chemistry and Physics*, 58th edn. (CRC Press, West Palm Beach, FL, 1997) F-210.
2. F. W. Sear and E. L. Sallinger, *Kinetic Theory and Statistical Thermodynamics*, Chap. 3.
3. G. W. C. Kaye and T. H. Laby, *Tables of Physical and Chemical Constants*, 16th edn. (Longman, 1995).
4. G. R. Fowler and G. L. Cassidy, *Analytical Mechanics*, 7th edn. (Thompson, 2005), Chap. 5.

Chapter 3

THE GLOBAL CLIMATE

3.1 Introduction

Although the global climate of the Earth is a complex function of many interacting variables, some appreciation of the factors that determine climate can be gained through the application of normal physical laws. This chapter seeks to provide a simple but hopefully not simplistic description of the global climate and its interactions in terms of the underlying physical mechanisms.

The principal variable commonly used in discussions of the global climate is the global temperature. This is understood as some mean value, averaged over the Earth and over time, since variations in solar irradiance and the heat carried by long-range ocean currents result in a non-uniform distribution of surface temperature. It is a matter of common experience that the temperature varies with both latitude and season. Equatorial temperatures are some tens of degrees higher than those at the poles, and even in the temperate mid-latitudes, there may be twenty to thirty degrees difference between summer and winter. Mean global temperature, therefore, is at best a very rough approximation and in reality not much more than a convenient index of the overall global state.

The Earth receives its energy primarily from the Sun in the form of radiation. In turn, the Earth re-radiates the energy back into space, indeed, if it did not, then the temperature would increase monotonically and without limit. While this is self-evidently the case, it is the interactions within the troposphere and the oceans that determine the surface temperature and which gives cause for concern over global warming, particularly as a result of anthropogenic influences.

To a first approximation, the emission spectrum of the Sun corresponds to that emitted from a black-body[a] at a temperature of about 5770 K.[1] Such a body at a temperature T K will radiate energy per unit surface area at a rate (I Wm^{-2}) according to the Stefan–Boltzmann Law:

$$I(T) = \sigma T^4 \tag{3.1}$$

where σ is the Stefan–Boltzmann constant (5.67×10^{-8} Wm^{-2}K^{-4}). With a radius of 6.96×10^8 m and taking the effective surface temperature to be 5770 K, the

---

[a]By definition, a black-body is one that absorbs all incident thermal radiation, irrespective of wavelength.

Sun emits about 3.85×10^{26} W of power. Assuming this is emitted isotropically then at the mean Sun–Earth distance of 1.49×10^{11} m, the calculated intensity is $1368 \, \text{Wm}^{-2}$. This is the mean value of the solar irradiance at the outer boundary of the Earth's atmosphere and is known as the solar constant, S.

3.2 Solar Spectrum

The spectral distribution of the radiation emitted by a black body is described by the Planck energy distribution which gives the energy density in the frequency interval ν and $\nu + d\nu$:

$$dU = \frac{8\pi h \nu^3}{c^3} \frac{1}{e^{(h\nu/k_B T)} - 1} d\nu \qquad (3.2)$$

where h, k_B and c are Planck's constant, Boltzmann's constant and the speed of light respectively. The peak emission is at a photon energy of $\sim$2.48 eV corresponding to a wavelength of $\sim$500 nm which is in the blue/green part of the spectrum. Figure 3.1 shows the spectrum of the sun at the edge of the Earth's atmosphere and at sea level.[2] In passing through the Earth's atmosphere, the solar spectrum is changed considerably as a result of elastic (Rayleigh) scattering and absorption.

In elastic scattering, the direction in which the photon is travelling is changed, but the photon energy is unaffected. If the wavelength of the radiation is large in comparison to the size of the scattering molecules or atoms, then the scattering cross section can be calculated from classical electromagnetic theory. The cross section

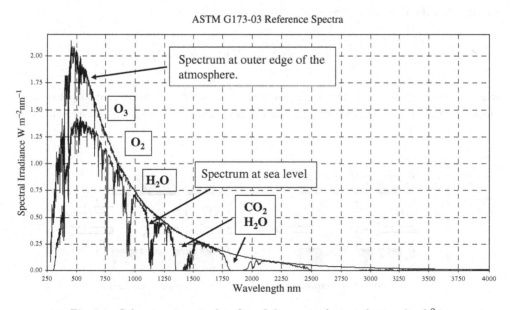

Fig. 3.1 Solar spectrum at the edge of the atmosphere and at sea level.[2]

was first worked out by Lord Rayleigh in 1871, who showed that the scattering probability is inversely proportional to the fourth power of the wavelength ($\propto \lambda^{-4}$). Consequently, blue light is more strongly scattered than yellow or red light and is the reason why the sky appears blue.

The sea level solar spectrum is also characterised by a number of absorption bands arising from absorption by gases in the atmosphere, principally H_2O, O_2, CO_2 and at the short wavelength end, ozone. The water vapour and CO_2 absorption bands are important contributions to global warming effects, while the ozone absorption provides protection from the harmful ultraviolet radiation. The strong H_2O absorption bands ($1.35 < \lambda < 1.45\,\mu$m; $1.8 < \lambda < 1.95\,\mu$m & $5.5\,\mu$m $< 7.5\,\mu$m) mean that the atmosphere is virtually opaque at these wavelengths, a fact of importance for infra-red sensor and night vision systems. In contrast, the atmosphere at sea level is quite transparent above this up to a wavelength of about $13\,\mu$m, a band often referred to as the $10\,\mu$m infra-red window.

Light is absorbed by a given atomic or molecular species if the photon energy corresponds to the difference in energy between some set of energy levels in the molecule or atom. On absorption, the photon excites an electron from the lower energy level E_1 to the upper level E_2, where the electron will remain for some characteristic average time before spontaneously relaxing back to E_1 (Fig. 3.2a). Alternatively, the electron may be stimulated into a downward transition by the presence of another photon of identical energy (Fig. 3.2b). The rates at which these three processes take place depend on the number of electrons (per unit volume) in the two levels, i.e. the population densities, N_1 and N_2, and the photon density, usually expressed as an energy density.

The rate of change in the population of E_1 when illuminated by light with an energy density $U(\nu)$ Jm^{-3} is:

$$\frac{dN_1}{dt} = A_{21}N_2 + B_{21}\,UN_2 - B_{12}\,UN_1 \tag{3.3}$$

where B_{12} and B_{21} are the Einstein coefficients for stimulated absorption and emission and A_{12} is the Einstein coefficient for spontaneous emission. In steady-state

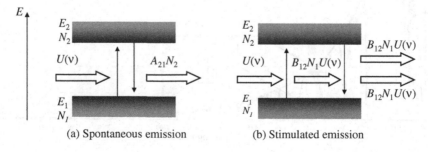

(a) Spontaneous emission   (b) Stimulated emission

Fig. 3.2   Optical absorption and emission processes: (a) spontaneous (b) stimulated.

when the rate of change is zero:

$$U(\nu) = \frac{A_{21}N_2}{B_{12}N_1 - B_{21}N_2} = \frac{A_{21}}{B_{12}(N_1/N_2) - B_{21}} \tag{3.4}$$

In thermal equilibrium at a temperature T the population ratio, N_1/N_2 is given by the Boltzmann distribution:

$$\frac{N_1}{N_2} = \exp\left(\frac{h\nu}{k_B T}\right). \tag{3.5}$$

Substituting into (3.4) and noting that the illuminating energy density, $U(\nu)$ is given by the Planck distribution, (3.2):

$$U(\nu) = \frac{A_{21}}{B_{12}\exp\left(\frac{h\nu}{k_B T}\right) - B_{21}} \equiv \frac{2\pi h\nu^3}{c^3\left(\exp\left(\frac{h\nu}{k_B T}\right) - 1\right)} \tag{3.6}$$

from which it follows that:

$$B_{12} = B_{21} \quad \text{and} \quad \frac{A_{21}}{B_{12}} = \frac{2\pi h\nu^3}{c^3}. \tag{3.7}$$

The stimulated transition rates (both absorption and emission) depend on the occupation of the two states in question, i.e. on N_1 and N_2, and since in general $N_1 \gg N_2$,[b] then a photon is much more likely to be absorbed than it is to stimulate an emission. Consequently, when light passes through some medium it will be attenuated. The degree of attenuation per unit distance is given by the Bouguert–Beer–Lambert Law (BBL) usually simply referred to as Beer's Law.

Consider the passage of light through a small slice of some medium, length dx and unit cross sectional area as illustrated in Fig. 3.3. As the light passes through the element some of it is absorbed with a transition rate between levels E_1 and

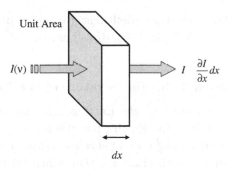

Fig. 3.3   Attenuation of light by an element of absorbing medium.

---

[b]It is only under conditions of very intense illumination where N_1 is depleted or in lasers where population densities are deliberately inverted that this is not the case.

E_2 given by (3.3), and assuming $N_1 \gg N_2$, and that the light intensity is not too strong (i.e. emission can be neglected)

$$\frac{dN_1}{dt} \approx -B_{12} N_1 U. \tag{3.8}$$

This must correspond to the rate of photon absorption, each of which contributes energy $h\nu$. The intensity at some photon frequency $I(\nu)$ is the amount of energy that flows through unit area in unit time, so the rate at which the intensity of the light is attenuated by absorption is

$$\frac{\partial I}{\partial t} = -B_{12} N_1 h\nu I. \tag{3.9}$$

It is more useful to know the rate of attenuation with path length, l rather than time, which may be obtained from (3.9) by the chain rule:

$$\frac{\partial I}{\partial x} = \frac{\partial I}{\partial t} \times \frac{\partial t}{\partial x} = -\frac{B_{12} N_1 h\nu}{(c/n)} I \tag{3.10}$$

where n is the refractive index (i.e. c/n is the velocity of light in the medium). Integration of (3.10) over the path length (from $x = 0$ to $x = l$) gives the BBL law:

$$I(l) = I(0) \exp(-Kl); \quad K = \frac{B_{12} N_1 h\nu}{(c/n)} = \frac{B_{12} N_1 hn}{\lambda}. \tag{3.11}$$

Conventionally, the BBL law is expressed as:

$$I(l) = I(0) \times 10^{-D}. \tag{3.12}$$

The index D is known as the optical density of the medium and is defined as:

$$D = \varepsilon(\lambda) Cl; \varepsilon(\lambda) = \frac{B_{12} hn N_A}{\lambda} \log(e) \tag{3.13}$$

where $\varepsilon(\lambda)$ is the molar extinction coefficient ($m^3 mol^{-1} m^{-1}$),[c] C is the molar concentration (mol m^{-3}) and N_A is Avogadro's number.

3.3 Radiative Balance and the Temperature of the Earth's Surface

The total solar power intercepted by the Earth is given by its cross sectional area πR_E^2 (where R_E is the radius of the Earth), but this is distributed over the surface of the globe (i.e. over an area $4\pi R_E^2$), and hence the *mean incident intensity* is $S/4$.

The main energy supply to the Earth is that which arrives from the Sun plus a small amount of geothermal energy coming from within due to the hot core of the earth. The solar energy input, the insolation, amounts to an average over the

---

[c]Frequently quoted in terms of $dm^3 mol^{-1} cm^{-1}$.

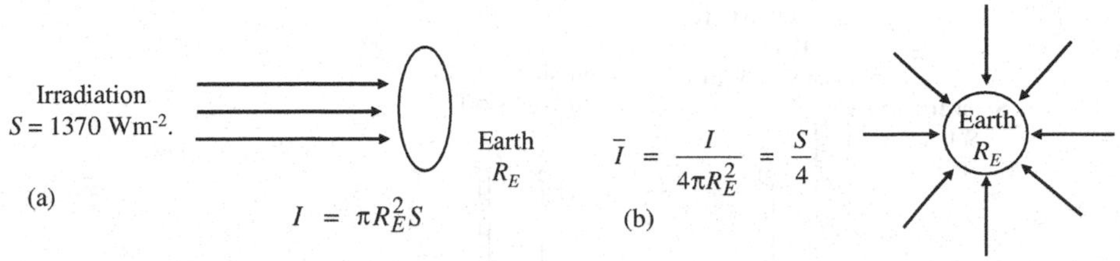

Fig. 3.4 Calculation of insolation: (a) intercepted radiation (b) average over surface.

surface at the outer edge of the atmosphere of about $342.5\,\mathrm{W\,m^{-2}}$ $(S/4)$. Of this, a fraction a_E, termed the albedo is reflected back from clouds, scattering by aerosols, etc. into space. The overall albedo has been estimated to be about 0.3, and thus the mean net insolation would, in the absence of atmospheric absorption, be about $0.7 \times (S/4)$ or $240\,\mathrm{Wm^{-2}}$. The geothermal contribution has been estimated to be about $0.082\,\mathrm{Wm^{-2}}$, negligible by comparison and will therefore be ignored in the following discussion.

In conditions of steady state, the energy received by the Earth from the Sun must be balanced by a corresponding loss of energy from the Earth by radiation into deep space. This is the principle of radiative balance:

$$\text{Energy in} = \text{Energy out}$$

or in terms of the Earth's temperature, T_{S}:

$$\frac{S}{4}(1 - a_F) = \sigma T_{\mathrm{S}}^4. \tag{3.14}$$

This gives an estimate for the mean temperature of the Earth of $255\,\mathrm{K}$, well below the freezing point of water and clearly much too low to sustain life. The actual average temperature for the Earth's surface is about $288\,\mathrm{K}$ ($15°\mathrm{C}$). Obviously, some places are much colder and some hotter, but since most of the planet is habitable the average temperature must be above the freezing point of water.

The reason for this discrepancy is the so-called Greenhouse effect due to differential absorption of incident and terrestrial radiation by atmospheric ('Greenhouse') gases, principally H_2O. The Earth receives short wavelength visible radiation from the sun, but re-emits at much longer infra-red wavelengths due to its lower temperature. The atmosphere as noted previously is almost opaque over much of the infra red spectrum and only a fraction of the radiation emitted by the surface of the Earth is transmitted through the atmosphere to outer space. The radiative balance is maintained by black body emission from the atmosphere. Figure 3.5 shows a much simplified schematic description of the processes.

Viewed from the perspective of outer space the incoming and outgoing radiated energy must balance, therefore

$$\frac{S}{4} = a_E \frac{S}{4} + \sigma T_S^4 + \sigma T_A^4 - \Delta I \tag{3.15}$$

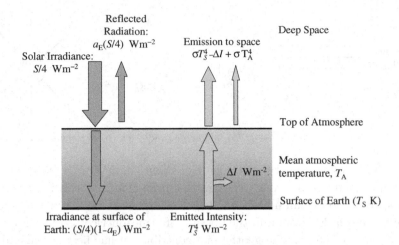

Fig. 3.5   Simple model of incident and emitted fluxes.

where T_A and ΔI are the mean temperature of the atmosphere and the emitted radiation absorbed by the atmosphere respectively.

3.4 Global Warming and Radiative Forcing

Absorption by greenhouse gases is essential if the Earth is to remain habitable, however, there is a balance to be preserved. If the concentration of greenhouse gases is too low, the surface temperature will drop and conversely, if the concentration is too high, the temperature will rise. Data compiled by the Intergovernmental Panel on Climate Control (IPCC)[3] demonstrates that there has been a steady increase in the level of CO_2 from $\sim$280 ppmv before the Industrial Revolution to a present day value of $\sim$370 ppmv — with almost half that increase having taken place within the past thirty years.

As more of the radiation emitted by the Earth's surface is absorbed by the atmosphere, the temperature of the Earth will be forced to increase in order to radiate the excess back into space, a process known as *radiative forcing*. Suppose that due to an increase in greenhouse gas concentration, an additional fraction ΔI is absorbed in the atmosphere rather than being radiated into space, then radiative forcing will result in a compensating increase ΔT_S in temperature. To first order, we can write

$$\Delta I = \frac{\partial I}{\partial T}\Delta T_S. \tag{3.16}$$

From the Stefan–Boltzmann Law (3.1) we get:

$$I = \sigma T_S^4; \quad \text{hence} \quad \frac{\partial I}{\partial T_S} = 4\sigma T_S^3 = \frac{4I}{T_S}. \tag{3.17}$$

The net insolation at the surface of the Earth is $(1 - a_E)S/4$, (see Fig. 3.5) and this must balance the emission from the surface and so replacing I in (3.17).

$$\Delta I = (1 - a_E)\frac{S}{T_S}\Delta T_S \quad \text{or} \quad \Delta T_S = G\Delta I; \quad G = \frac{T_S}{(1 - a_E)S} \tag{3.18}$$

where G (W^{-1} m^2 K) is the 'gain function'.

The Earth has a finite heat capacity and therefore any increase in absorbed energy, ΔI, will take time to raise the temperature. Equation (3.18) really only describes the situation after a steady state has been established. The additional energy absorbed has consequently to power the rate of heating and ultimately maintain the steady state.

$$\Delta I(t) = \frac{\Delta T_S(t)}{G} + c_E \frac{d}{dt}(\Delta T_S(t)) \tag{3.19}$$

where c_E is the 'mean' heat capacity of per unit area of the Earth's surface ($\text{Jm}^{-2}\text{K}^{-1}$). The first term on the right hand side is simply the energy required to maintain the surface temperature; the second term is the heat per unit time required to raise the temperature of a unit area of the Earth's surface at a given rate. The solution to (3.19) and hence the manner in which ΔT_S changes with time, depends of course on the nature of the increase in ΔI, but for a step change in absorbed radiation ($t < 0$, $\Delta I = 0$; $t > 0$, $\Delta I = \text{constant}(\neq 0)$), the increase in temperature follows a simple time constant behaviour of the form:

$$\Delta T_S = G\Delta I \left(1 - \exp\left(\frac{-t}{\tau}\right)\right) \tag{3.20}$$

where $\tau(= c_E G)$ is the time constant. Evidently, as time passes, the second term in the brackets diminishes and ΔT_S rises asymptotically to the new level required to sustain radiative balance.

By the same token, any reduction in ΔI (e.g. by reducing CO_2 levels) will not lead to an immediate drop in the temperature — compounding delays in remedial action to limit anthropogenic sources of greenhouse gases.

3.5 Feedback Effects

Implicit in the derivation of the gain function (3.18) is the assumption that the resulting change in temperature does not produce any effects that either reinforce or alternatively oppose the change, i.e. there are no feedback effects. This is clearly unrealistic and in practice there are many feedback effects that need to be taken into account. Increases in global temperature, for example, may be expected to result in the melting of polar ice caps with a consequential reduction in albedo and an increase in the gain. Conversely, desertification acts to increase the albedo (sand has a higher albedo than foliage) and hence lower the gain.

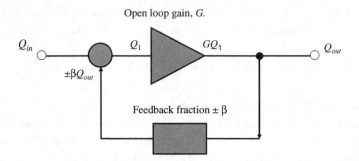

Fig. 3.6   Feedback principles.

Feedback principles are well established and commonly used to control the performance of machinery and instrumentation. The principle is illustrated in Fig. 3.6. Some fraction (β) of the system's output is fed back so as to either add to (positive feedback) or subtract from (negative feedback) the input, depending on the sign of β.

The parameter of interest is the overall gain including feedback effects, G_f, defined as the ratio of the final output (Q_{out}) to the input (Q_{in}). With reference to Fig. 3.6,

$$Q_{\text{out}} = GQ_1 = G(Q_{\text{in}} + \beta Q_{\text{out}}) \tag{3.20}$$

and thus there is the overall gain with feedback;

$$G_f = \frac{Q_{\text{out}}}{Q_{\text{in}}} = \frac{G}{1 - G\beta}. \tag{3.21}$$

Equation (3.21) is known as the 'general equation of feedback' and, with the appropriate parameters, applies to any feedback process. If β is positive (positive feedback), then the feedback acts to increase the overall gain. This is an unstable situation which can lead to a runaway situation. In particular, if the product (known as the loop gain) $G\beta = 1$, then in principle G_f becomes infinite! (Actually, this is the condition for an oscillator — i.e. an output with no input). Conversely, negative feedback ($\beta < 0$) reduces the overall gain and hence increases stability.

In the present case, there are many factors contributing to the feedback, some additive, others subtractive and not all of them independent of each other. The feedback fraction is therefore a function of all these influences and we have to use an effective feedback fraction, β_{ef}, in (3.21).

$$G_f = \frac{G}{1 - G\beta_{ef}}; \quad \beta_{ef} = f(\beta_i) \tag{3.22}$$

where β_i represent all the individual feedback effects (note that some of the β_i will also in general be functions of several interacting processes).

What is of primary concern is the sign of β_{ef}, in other words, whether the overall feedback is positive and the warming processes are therefore unstable, or negative and reasonably well-damped. This is not a simple question to answer and much current research is focused on establishing the sign and magnitude of the feedback.

Identifying whether some consequence of global warming has a positive or negative feedback or is neutral is in general not straightforward, as many have both benign and malign impacts on global warming. Determining the extent to which such influences cancel each other is difficult and probably dependent on the degree of global warming. For example, the solubility of CO_2 in the oceans decreases as the temperature increases, and thus warmer oceans, we might suppose, would release more CO_2 into the atmosphere and through radiative forcing act to increase global warming, i.e. positive feedback. However, warmer ocean temperatures would stimulate algae growth, in turn consuming more CO_2. The situation is even more complicated, because even if these two opposing effects could be considered in isolation (which they probably cannot), they will be characterised by different time constants. With these caveats, a number of possible feedback effects are summarised in Table 3.1.

The impact of changes in lapse rate, Γ (see Sec. 2.3.1), associated with changes in humidity are not immediately obvious, but considerable. Lapse rate is the rate at which the temperature of a rising parcel of air changes with altitude in the lowest few km of the atmosphere. Suppose there is an increase in the concentration of CO_2 in the atmosphere, long wavelength absorption increases and due to radiative forcing, so does the surface temperature of the Earth. In dry air conditions, the lapse rate remains essentially unchanged at $\sim 10^{-2}$ K m^{-1} and there is no additional warming or cooling effect. However, the lapse rate of warm, humid air is smaller than for dry (or cool air) and so at a given altitude, the temperature in moist air will be greater than it would be at the same height for dry air. In consequence, the emission from the atmosphere can partly compensate radiative forcing and thus result in a net cooling effect.

Table 3.1   Some positive and negative feedback contributions to global warming.

Positive Feedback Effects
1.   Melting of the polar ice caps — reduces albedo;
2.   Warmer oceans dissolve less CO_2 — increases release of CO_2 into the atmosphere;
3.   Warmer oceans result in greater evaporation — H_2O is a good greenhouse gas;
4.   Condensation of water vapour in clouds releases latent heat of condensation — warming effect;
5.   Warmer temperatures stimulate greater plant growth — reducing albedo.

Negative Feedback Effects
1.   Warmer oceans promote algae growth — increases consumption of atmospheric CO_2;
2.   Evaporation from the oceans absorbs latent heat of evaporation — cooling effect;
3.   Burning of fossil fuels releases aerosols — scattering predominates over absorption, increasing albedo;
4.   Increased plant growth sequestrates more CO_2 — reducing CO_2 concentration in the atmosphere;
5.   Higher evaporation from the oceans — reduces lapse rate.

3.6 The Role of Carbon Dioxide

Absorption of a photon of radiation by a molecule causes the atoms within that molecule to vibrate in specific ways and frequencies that are unique to that particular molecule. Only photons with energies that match the excitation energies of the individual vibrations will be absorbed. This is the basis of molecular spectroscopy.

In the case of carbon dioxide in the ground (i.e. unexcited) state, the oxygen atoms are located on either side of the central carbon atom so that all three are colinear. There are, in principle, three ways or *normal modes* in which a CO_2 molecule can vibrate (see Fig. 3.7):

(a) the symmetric stretch mode, where the two oxygen atoms oscillate *along* the molecular axis such that they are always equidistant from the carbon atom (the O atoms move in *anti-phase* to each other);

(b) the asymmetric stretch mode, where the oxygen atoms oscillate along the molecular axis, such that they are not in general equidistant from the carbon (the O atoms move *in phase* with each other);

(c) the bending mode, where the oxygen atoms oscillate along arcs centred on the carbon atom and in anti-phase with each other (rather like the flapping of a bird's wings).

Selection rules for infrared activity require that the motion corresponding to the active mode should result in a change in the dipole moment for the molecule. Of the three normal modes, only (b) and (c) involve a change in the dipole moment and hence are infrared active. The bending mode has the lowest fundamental wavelength corresponding to about $15\,\mu$m. The asymmetric stretch mode has a much shorter fundamental wavelength[4] of $4.9\,\mu$m. The bending modes are more easily excited as this mode is in local thermodynamic equilibrium with the atmosphere, by which

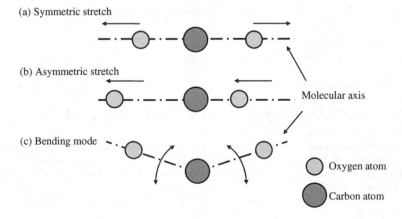

(a) Symmetric stretch

(b) Asymmetric stretch

Molecular axis

(c) Bending mode

Oxygen atom

Carbon atom

Fig. 3.7   Normal vibrational modes for the CO_2 molecule.

we mean that relative populations are well described by the Boltzmann distribution (3.5) and coupling of the gas to the radiation field is strong. (The shorter wavelength asymmetric vibrations are not in local thermodynamic equilibrium and are hence not so readily excited — the system is said to be in vibrational relaxation). The principal absorption band is therefore that centred at a wavelength of 15 μm, close to the peak emission from the Earth of about 10 μm, assuming a mean temperature of 288 K. Consequently, the shorter wavelength incoming solar radiation is not strongly absorbed by CO_2, but the outgoing longer wavelength is and gives rise to the greenhouse effect.

Carbon dioxide is by no means the only greenhouse gas nor is it the strongest: the global warming effect per molecule of methane, for example, is some twenty times that of CO_2. The importance of CO_2 arises from its abundance in the troposphere and the clearly established observational fact that the atmospheric concentration is increasing monotonically. To provide some means of comparison between different greenhouse gases and their inclusion in calculations we define the 'Global Warming Potential' of a gas as the estimated effect 100 years after releasing 1 kg of the gas relative to the release of 1 kg of CO_2. Gases with a long residence time in the atmosphere are correspondingly more significant than gases with short residence times. These calculations show that methane (which is produced by the decay of vegetable matter) has a global warming potential of $\sim$21 while nitrous oxide (N_2O) has a potential of 310. (By definition, CO_2 has a potential of 1).

3.7 Climate Variations

The Earth's climate is a dynamic system that was subject to variation long before the advent of humanity, as witnessed by the ice ages. Studies of pollen data, deep sea sediments and ice cores from Greenland have enabled estimates to be made of the variation in air temperature and by inference, surface temperature over the past million years or so. The ice core data relies on the precise measurement of the relative ratios of the oxygen isotopes ^{16}O and ^{18}O in the ice. The underlying principle is that the heavier ^{18}O evaporates more slowly from the oceans than the lighter ^{16}O and precipitates out as rain more readily. Consequently, by the time the water vapour has reached the polar regions, the isotope ratio would have changed. The actual ratio in the Greenland ice core will depend on the temperature in the equatorial regions where most of the evaporation was assumed to occur, precipitation en route and the air temperature in Greenland when the remaining water vapour fell as snow. Empirical calibration of the process allows the temperature at the time the core was formed to be estimated with a surprising degree of accuracy. Complementary information may be obtained from the deuterium content and sodium content. Small entrapped gas bubbles in the ice cores provide an indication of the atmospheric levels of CO_2 and CH_4.

The results show that the mean global temperatures varied by up to $\pm 4°C$, periods of low temperature obviously corresponding to ice ages.[5] The data suggest that the long term mean global temperature over the last million years has been about 13°C, but that over the last 10,000 years it has been nearer 15°C.

In 1998, ice core drilling at the Vostok[d] project produced the deepest ice core recovered to date of 3,623 m. This corresponds to a period of 420,000 years and covers the past four glacial-interglacial cycles. Analysis of the core has shown that the atmospheric concentration of CO_2 (and (CH_4)) correlate well with Antarctic air temperature.[5] The study also showed that while the atmospheric properties oscillated within stable bounds, present day concentrations of CO_2 have "been unprecedented during the past 420,000 years."

Following the work of the astronomer Milankovitch,[6] the variations are generally attributed to differences in insolation due to small changes in the Earth's orbit and inclination. The Earth follows an elliptical orbit with the Sun at one focus, and due to gravitational interactions with the rest of the solar system, the eccentricity varies with a period of about 100,000 years (Fig. 3.8(a)).

The rotational axis of the Earth is tilted with respect to the normal on the orbital plane by between 21.5° and 24.5°, and varies with a period of about 41,000 years (Fig. 3.8(b)). The Earth is slightly oblate (equatorial diameter > polar diameter) and the torque exerted by the moon on the spinning oblate causes the axis to precess with a period of about 26,000 years (Fig. 3.8(c)). These three independent periodicities will interfere with each other to cause long-term changes in the global climate.

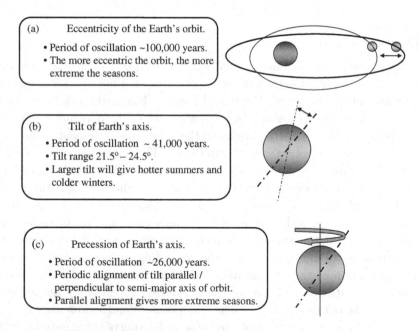

Fig. 3.8   Contributing factors to the Milankovitch model.

---

[d]An international project involving teams from Russia, USA and France.

Seasonal variations in insolation will be greater when the Earth is in a more eccentric orbit and in consequence, the differences between the two hemispheres will be more significant. The hemisphere for which summer occurs when the Earth is nearest the Sun will have hotter summers but correspondingly colder winters, while the converse will be true for the other hemisphere. Larger tilt angles will cause more extreme conditions for both hemispheres, while the precession will determine at what point in the orbit the summer and winter solstices occur and the amount of daylight.

The rates of change associated with the long term variations in the Earth's orbit and inclination are relatively slow, of the order of 0.1°C per 1000 years, much less than changes measured over recorded history. Generally, this disparity is assumed to be the result of non-linearity in the equations of motion, implying that the behaviour is chaotic.

3.8 The Impact of Global Climate Change

The implications of changes in global climate are very difficult to predict. Many of the consequences are interdependent and chaotic in character, requiring extensive computational facilities and sophisticated techniques to model. Nevertheless, two consistent themes appear to be emerging from most of the studies being conducted throughout the world. Firstly, effects will be regional, and secondly, warming will cause significant rises in sea level. Although time scales differ from model to model, the general conclusions are much the same.

Warming will, not surprisingly, lead to increased melting of the polar ice caps, resulting in turn to a reduction in the albedo and further warming: a positive feedback effect. The progressive release of water presently locked up in the ice fields and glaciers of Antarctica and the northern hemisphere tundra is predicted to raise sea levels by as much as 2 mm per annum, threatening the survival of many of the world's major cities, most of which were built on coastal sites as trading ports. Flooding of coastal plains will inevitably have demographic implications as populations will have to move inland increasing stress on land usage. Some low lying Pacific islands may be engulfed completely.

Less obvious is the impact that melt water from the polar ice fields may have on the great ocean currents. The Gulf Stream, for example, is crucial in maintaining the climate of maritime Western Europe warmer than would be expected given its latitude. The Gulf Stream conveys shallow warm water northwards from the confluence of warm currents flowing up the west coast of Africa and the east coast of South America. In due course, the stream is cooled by the arctic polar air mass and as a result, its density increases and it sinks to the considerable depths (~4000 m) of the North Atlantic Deep Water. The continually descending water causes the cold deep water to spread out southwards, eventually reaching the southern tip of Africa. There it is joined by cold water from the Antarctic and gradually drifts northwards into the Indian and Pacific oceans, where ultimately, the water starts to warm again and rise to the surface. The warmer surface water returns to the Atlantic to repeat the cycle. The ocean currents are driven by the thermohaline cycle; the impact on the density of the sea water from salinity and temperature. The release of large volumes of fresh, less dense melt water into the Northern Atlantic would shift the cycle southwards. The warmer, more saline water would sink at lower latitudes potentially before reaching European coastal regions. Ironically, one consequence of global warming might be cooling in Western Europe.

Regional climatic changes will become manifested by changes in rainfall patterns; some areas becoming drier and others less so, with obvious implications for agriculture. The difference in temperature between the equator and the poles is one of the main drivers of the atmospheric global circulation. Global warming is expected to be more pronounced at the poles (due to the locally reduced albedo) than at the equator, and the resulting changes in circulation will produce corresponding changes in rainfall distribution.

Another consequence of global warming is variability and change in weather. The most dramatic examples of this have been the increased incidence of El Niño events and the increasing severity of hurricane storms. Historically, El Niño (Spanish for a male child) events occurred every three to seven years and were associated with unseasonable and often disastrous weather conditions throughout the southern hemisphere. The genesis of El Niño incidents is still not fully understood, but they are thought to result from a reduction in strength of the Easterly Trade Winds. In non-El Niño years (Fig. 3.9), the Easterly Trade winds blowing off the western coasts of South America 'drag' warm surface water from the coasts of Peru and Ecuador raising the sea level on the Western Pacific (Indonesian coast) by up to half a metre. The Western Pacific is therefore much warmer and the atmosphere above it becomes laden with moisture and becomes unstable, with as a result heavy rainfall in the area, i.e. the Philippines, Southern India and Sri Lanka etc. The easterly current flowing away from the western seaboard of South America is replenished by upwelling cold water leading to the opposite effects. The atmosphere is cooled and becomes stable with little rainfall producing the coastal deserts of the region.

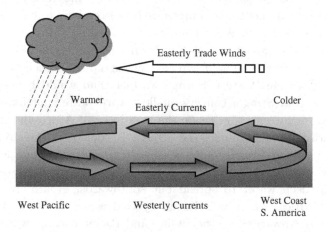

Fig. 3.9   El Niño sequence.

For reasons that are not fully understood, in El Niño years, the pressure gradient which drives the Easterly Trade winds is reduced, lessening the strength of the winds. In turn, the strength of the Easterly current is decreased and thus the western Pacific is cooler and its atmosphere is more stable, while the Eastern side is warmer and becomes less stable. Drought conditions then prevail in the normally wet countries of Indonesia, the Philippines, Southern India and Sri Lanka, and conversely, there is heavy rainfall and flooding in Bolivia, Ecuador, Peru and the US Gulf states. El Niño events have become more frequent (every two to three years) and more intense, although whether El Niño events are a consequence of climate change has yet to be demonstrated. The 1997/98 El Niño, which was particularly severe and destructive, began in the spring of 1997 when ocean borne sensors recorded abnormally high temperatures. Over the following months, as the El Niño matured, the temperature increases rose to a 50 year high (i.e. since the advent of reliably recorded data) of 4°C. This corresponded to extreme drought and extensive forest fires in Indonesia and the Philippines, and severe flooding in South American States. It has been estimated that up to 2,000 deaths were attributable to this particular El Niño.

Apart from the climatic and weather aspects, the El Niño has major economic implications. The upwelling cold water brings high levels of nutrients, supporting large populations of fish and sustaining a substantial fishing industry. When the upwelling is less pronounced, nutrient levels are lower and become rapidly depleted leading to an equally rapid decline in fish stocks[e].

Recent research has indicated that although the frequency of hurricanes in the eastern Atlantic has remained much the same, the severity of the storms has increased. This would be consistent with an increase in the mean water temperature of the western equatorial Atlantic where the storms form (see Sec. 2.4.6).

3.9 The International Politics of Global Climate Change: Rio de Janeiro and Kyoto

Although technological improvements in power stations, car and plane engines, the deployment of a larger fraction of renewable energy etc. can all help reduce the emission of greenhouse gases, such measures on their own are unlikely to prove sufficient without consensus among the international community to take meaningful steps to limit the use of fossil fuels. This and other ecological issues of concern prompted the UN to call the first Earth Summit in Rio de Janeiro, in 1992.

Greenhouse gas emission was high on the agenda of the Rio summit and resulted in the *Framework Convention on Climate Control* (FCCC), one of five documents signed by the participating Heads of States. The FCCC had the modest aim of cutting back emissions to 1990 levels by the year 2000 (an indication

---

[e]Traditionally, the first sign of an El Niño event is the appearance of large numbers of dead fish littering the beaches.

of how rapidly the levels were rising) and signatory countries agreed voluntarily
to take the necessary steps to do so. Within five years, it was apparent that
greenhouse gas emissions were continuing to rise inexorably and that the voluntary
basis of the FCCC was not working. With the exception of the European Union,[f]
developed countries increased their emissions by 7% to 9%, while developing
countries increased theirs by as much as 25%, though from a lower base level.

This led to a meeting in December 1997 in Kyoto, intended to draft a more
binding treaty on FCCC signatories, the *Kyoto Protocol*. The developing countries
argued successfully that since the industrialised nations were to blame for current
levels of CO_2 in the atmosphere, they should take the 'pain' — 'differential
responsibilities' as the policy is now termed. In response, industrialised countries
accepted the argument and committed themselves to cutting back emissions to pre-
1990 levels by 2010. The Kyoto Protocol placed an agreed limit on the greenhouse
gas emissions of each country. Under the rules, the protocol would only come
into force if it was ratified by the legislatures of countries responsible for at least
55% of emissions. The agreement ran into immediate problems in the USA, the
world's largest emitter of CO_2, when the Senate, under pressure from oil and
industrial interests and fearful of competition from India and China, refused to
ratify the treaty, as required by the US constitution. This meant that nearly all
the other industrial nations would have to ratify the agreement for it to become
effective. This was eventually achieved when Russia ratified the treaty and the
Kyoto Protocol officially came into force on 16th February 2005. The challenge for
developed countries that ratified the Kyoto Protocol is how to achieve the cutback
targets.

The UK government agreed to a 12.5% reduction, predicated firstly on a switch
from coal to gas (i.e. North Sea gas) fired power stations for the production of
electricity (modern gas fired stations are much more efficient than older coal fired
systems — see Sec. 6.5) and secondly, on a reduction in traffic consumption.
Unfortunately, the first part of this strategy fell foul of rises in the price of gas,
as the North Sea fields started to run out, causing some electricity suppliers to
switch more of their production back to coal.

The reduction in traffic consumption was to be achieved by progressive increases
in the rate of taxation — the fuel escalator — in the expectation that people would
use their cars less and choose smaller, more fuel efficient models. In any event, the
fuel escalator proved so unpopular that it had to be quickly withdrawn. The UK
government also promoted renewable energy sources leading to the proliferation of
(equally unpopular) wind farms, from which the power utilities are compelled to
buy more expensive 'green' electricity. In spite of these difficulties, the UK is still
below its target of 12.5%, although it may become increasingly difficult to remain so

---

[f]Within the E.U. there were wide disparities: in some countries such as Ireland, emissions increased
dramatically, while in others such as the UK and Germany they declined — the net result was a
reduction.

in the future. Anticipated increases in demand are expected to outstrip the capacity of renewable sources, resurrecting the debate over nuclear power.

Apart from measures to reduce carbon emissions at home, countries can 'trade' their emissions on the basis that the atmosphere does not 'care' where the emissions come from. This can be done in three ways:

(i)   Joint implementation schemes: If a country, such as Britain for example, were to spend money making a power station in another developed country e.g. Poland, more efficient than if the same investment were made in Britain making an already efficient power station slightly more efficient, then both countries can claim a share of the carbon saving;

(ii) Clean development mechanism: If instead the investment were made in installing a completely new and carbon-free power station (e.g. solar powered) in a developing country, then all the carbon saving is attributed to the donor country;

(iii) Carbon trading: This is where developed countries that have ratified the Protocol can purchase some of the carbon savings made by a third country. This is particularly attractive to the former communist block countries, which because of the closure of their heavy industries have saved substantial quantities of carbon emissions, but are in need of cash.

One criticism of (iii) is that it does not directly produce a reduction in CO_2 emissions, although the expense should act as a disincentive for the purchasing country to rely on carbon trading unduly.

The Kyoto Protocol is a legal agreement with penalties for failure to meet target reductions. Countries that do not meet their reductions by 2010 will be called to account for it. If judged to have been cavalier, they could face a number of penalties, including exclusion from the three forms of trading agreements listed above. In addition, they could have the shortfall increased by a third, and added to their target for the following period.

3.10 Problems

1. Show that for an ideal gas at constant temperature, the optical density (D) is directly proportional to the pressure of the gas.

    The intensity of light transmitted through a 5 cm long test cell is reduced by a factor of 0.8 when it is filled with a particular gas at atmospheric pressure compared to the intensity measured when the cell is evacuated. If the temperature was 288 K, estimate the molar extinction coefficient of the gas.

    (Universal Gas Constant, $R = 8.314 \, \mathrm{J \, mol^1 \, K^{-1}}$).

2. Show that if there were no global warming at all, the average surface temperature of the Earth would only rise just above the freezing point of water, even if the albedo were zero.

What value of albedo would reduce the mean surface temperature to the freezing point of water?

(take solar constant, $S = 1370\,\mathrm{Wm^{-2}}$)

3. Some models of global warming imply that the rate of change in temperature, T with carbon dioxide concentration, C is inversely proportional to C, i.e.

$$\frac{dT}{dC} \propto \frac{1}{C}.$$

It is generally accepted that a doubling of the CO_2 concentration in the atmosphere results in an increased absorption of infrared radiation from the Earth of $\sim 4.6\,\mathrm{Wm^{-2}}$.

At the end of the 20th century, the mean CO_2 concentration in the atmosphere was $\sim 370\,\mathrm{ppmv}$. In the 'business as usual' scenario, the CO_2 concentration is projected to rise to $\sim 550\,\mathrm{ppmv}$ by the year 2030.

Ignoring feedback effects, what is the expected rise in mean temperature between the end of the 20th Century (mean temperature 288 K) and the year 2030, if the albedo remains constant at 0.3?

(take the solar constant to be $1370\,\mathrm{Wm^{-2}}$).

4. What would be the expected temperature increase in question 3, if there was positive feedback with a feedback fraction (β) of 0.9?

References

1. G. Woan, *The Cambridge Handbook of Physics Formulas* (Cambridge University Press, 2003).
2. http://rredc.nrel.gov/solar/spectra/1.5; NREL ASTM C173-03 Reference Spectra.
3. J. T. Houghton, G. J. Jenkins and J. J. Ephraums (eds.), *Climate Change: The IPCC Scientific Assessment* (Cambridge University Press, 1990).
4. J. W. Chamberlain, Theory of Planetary Atmospheres. An Introduction to their Physics and Chemistry, *Int. Geophysics Ser.* **22** (1978).
5. J. R. Petit, J. Jonzel, D. Raynard, N. I. Barkov, J.-M. Barnola, I. K. Basile, M. Bender, J. Chapallaz, M. Daris, G. Deaygue, M. Delmotte, V. M. Kotlyakov, M. Legrand, Y. V. Lipenkov, G. Lorius, L. Pépin, C. Ritz, E. Saltzman and M. Stievenard, Climate and Atmospheric History of the Past 420,000 Years from the Vostok Ice Core, Antarctica, *Nature* **399** (1999) 424–436.
6. M. Milankovitch, *Kanon der Erdbestrahlungen und seine Anwendung uf das Eiszeitenproblem* (Belgrad, 1941). (Canon of Insolation and the Ice Age Problem, English translation Alve. Global, 1998).

Chapter 4

SOLAR ULTRAVIOLET RADIATION AND LIFE

4.1 Solar Ultraviolet Spectrum

As noted in the previous chapter, the solar spectrum at the outer limit of the Earth's atmosphere is essentially that of a black body at a temperature of 5770 K (see Sec. 3.1). Although the peak emission is in the blue/green part of the spectrum ($\lambda \sim 500$ nm), a considerable fraction of the solar irradiance lies at shorter wavelengths, and in particular, in the ultraviolet (UV) ($\lambda < \sim 480$ nm). However, most of the short wavelength UV is absorbed in the upper atmosphere (stratosphere and above), principally by ozone (O_3) and the spectrum observed at sea level is therefore truncated at the short wavelength end (see Fig. 3.1). Sea level irradiance is effectively zero at a wavelength of about 290 nm, although the corresponding irradiance at the edge of the atmosphere is still considerable ($\sim 10^3$ times larger).

Conventionally, the UV part of the solar spectrum is divided into three bands, broadly associated with their biological significance (Fig. 4.1). The near UV part (320 nm $< \lambda < 400$ nm) is termed the UV-A and is not absorbed by biological molecules. Absorption in the far UV or UV-C (200 nm $< \lambda < 290$ nm) is intense, but this band lies beyond the sea-level spectrum and in everyday health terms is also of little significance. The most dangerous part of the solar spectrum at sea level are the intervening wavelengths, the UV-B band (290 nm $< \lambda < 320$ nm) where the absorption bands of important biological molecules overlap with the solar spectrum.

The harmful effects of exposure to UV-B (and UV-C) are not confined to animals but to plants and single-celled life forms as well, since the damage is inflicted on the DNA molecules common to all forms of life. The natural cycle of ozone creation and annihilation which provides the UV shield is therefore crucial to the survival of life as we know it. It is clearly important that the absorption takes place well above the biosphere i.e. in the stratosphere. The density is very low at such altitudes and in reality, there is little ozone present, its concentration is therefore particularly sensitive to the mechanisms of depletion, especially those associated with anthropogenic sources. Concern over ozone depletion led to the Montreal Protocols, the first, and at the time of writing, still the most successful international agreement on matters of the environment.

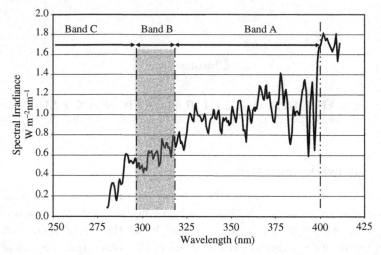

Fig. 4.1 Expanded UV part of the extra-terrestrial solar spectrum delineating the UV-A, UV-B and UV-C bands. The highlighted section corresponds to the dangerous UV-B band.[1]

4.2 The Ozone Filter

4.2.1 *The Chapman reactions*

Ozone absorption takes place in the stratosphere by a self sustaining natural cycle of UV photolytic and chemical reactions between the three main allotropes of oxygen (O, O_2 and O_3). The photochemistry was first described by Chapman in a series of papers in 1930 and 1931[2-4] and is embodied in the sequence of reactions that have since come to be known by his name.

Ozone is created in a two stage process (Fig. 4.2(a)), involving firstly the photolytic dissociation of an oxygen molecule into two individual atoms by an

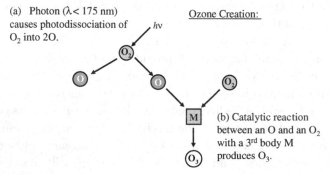

Fig. 4.2 Chapman reactions for ozone creation: (a) UV photolytic disassociation of O_2 yields two oxygen atoms (b) catalytic reaction of O and O_2 in the presence of an unspecified body M produces the ozone.

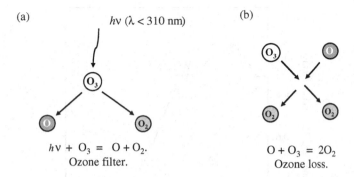

Fig. 4.3 Chapman reactions for ozone destruction: (a) UV photolysis of O_3 to O_2 and O provides the UV screen (b) ozone loss due to reaction of O_3 and O to form two O_2 molecules.

energetic UV photon ($h\nu > 7\,\text{eV}$, $\lambda < 175\,\text{nm}$)

$$O_2 + h\nu \rightarrow 2O. \tag{4.1}$$

As a consequence, there is total extinction of incident solar UV radiation at wavelengths shorter than this. One of the O atoms then combines with an oxygen molecule in the presence of an unspecified catalytic third body M (e.g. an aerosol particle) to form the ozone molecule (Fig. 4.2(b)).

$$O + O_2 + M \rightarrow O_3 + M. \tag{4.2}$$

Ozone is converted back to O and O_2 by the dissociative absorption of UV light of wavelengths below $\sim 310\,\text{nm}$ ($h\nu > 4.1\,\text{eV}$) as illustrated in Fig. 4.3(a)).

$$O_3 + h\nu \rightarrow O + O_2. \tag{4.3}$$

Both the reaction products are available to form another molecule of ozone by reaction (4.2). It is the combined effects of reactions (4.1) to (4.3) that constitute the UV filter.

Monatomic oxygen and ozone may also be destroyed by direct reaction between them to form two molecules of oxygen (Fig. 4.3(b)).

$$O + O_3 \rightarrow 2O_2. \tag{4.4}$$

Reaction (4.4) does not involve any photolytic reactions (i.e. no UV screening) and therefore corresponds to a net loss of O_3. However, as the concentrations of O and O_3 are extremely small, the impact of this process is not too serious.

The concentration of O_3 in the stratosphere is rather low ($< 5 \times 10^6$ molecules m^{-3}), but this is offset to some degree by the large absorption cross section ($> 10^{-22}\,\text{m}^2$) in the Hartley absorption continuum[5] (Fig. 4.4) which peaks at $\lambda \sim 255\,\text{nm}$.

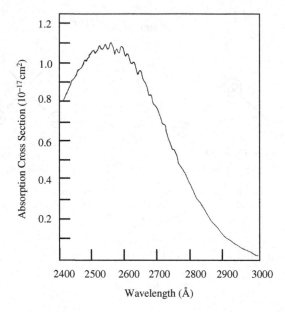

Fig. 4.4   The Hartley continuum for ozone absorption.[5]

4.2.2 *Reaction rates*

There will obviously be a diurnal variation in O_3 abundance as reactions (4.1) and (4.3) can only proceed in daylight. Suppose reaction (4.1) proceeds at a rate of J_1 per second per molecule of O_2 (i.e. the photolysis rate) which depends on the photon flux and the cross section for absorption, and generates two atoms of oxygen (i.e. generation rate of O per molecule of O_2 is $2J_1$). Similarly, let reaction (4.3) proceed at a rate J_3 ($s^{-1}O_3$ molecule^{-1}) destroying one molecule of O_3. The other two reactions (4.2) and (4.3) are characterised by rate coefficients k_{12} (m^6s^{-1}) and k_{13} (m^3s^{-1}).

Monatomic oxygen is involved in all four reactions, (4.1) and (4.3) increase the concentration of O, while (4.2) and (4.4) reduce it. Denoting the concentrations by square brackets, the net rate of change of monatomic oxygen is obtained by summing the individual rates

$$\frac{d[O]}{dt} = 2J_1[O_2] + J_3[O_3] - k_{13}[O][O_3] - k_{12}[O][O_2][M]. \qquad (4.5)$$

Ozone is involved in three of the reactions, two of them, (4.3) and (4.4), are destructive and one creative, (4.2). In the same way we can write for the rate of change O_3 with time

$$\frac{d[O_3]}{dt} = k_{12}[O][O_2][M] - k_{13}[O][O_3] - J_3[O_3]. \qquad (4.6)$$

In the steady state, the concentrations remain constant and the left hand sides of both (4.5) and (4.6) vanish. The steady state photolysis rate for the dissociation of O_3 is from (4.6)

$$J_3[O_3] = k_{12}[O][O_2][M] - k_{13}[O][O_3] \tag{4.7}$$

and substituting this in (4.5) we get for the daytime steady state concentration of [O]

$$[O]_{\text{Day}} = \frac{J_1[O_2]}{k_{13}[O_3]}. \tag{4.8}$$

Collecting terms in (4.7) gives the steady state concentration of O_3

$$[O_3]_{\text{Day}} = \frac{k_{12}[O][O_2][M]}{k_{13}[O] + J_3} \approx \frac{k_{12}[O][O_2][M]}{J_3} \tag{4.9}$$

where we have made use of the observation that below $\sim 60\,\text{km}$, the abundance of monatomic oxygen is low and $k_{13}[O] \ll J_3$.

Eliminating [O] from (4.9) by substitution from (4.8)

$$[O_3]_{\text{Day}} \approx \left\{ \frac{k_{12}}{k_{13}} \frac{J_2}{J_3} \right\}^{\frac{1}{2}} [M]^{\frac{1}{2}} [O_2] \tag{4.10}$$

from which it is apparent that the concentration of ozone is proportional to that of molecular oxygen. The ozone concentration is also proportional to the square root of the oxygen photolysis rate (J_2) and inversely to the square root of the ozone photolysis rate (J_3). Photolysis rates depend on the spectrum of the incident light and since this changes with altitude, the ratio J_2/J_3 is not constant but also varies with altitude.

Clearly, at night both J_2 and J_3 are zero and only reactions (4.2) and (4.4) can occur. The rates of change in [O] and [O_3] now become:

$$\left. \frac{d[O]}{dt} \right|_{\text{Night}} = -k_{12}[O][O_2][M] - k_{13}[O][O_3] \tag{4.11}$$

which is negative (i.e. [O] decreasing) and

$$\left. \frac{d[O_3]}{dt} \right|_{\text{Night}} = k_{12}[O][O_2][M] - k_{13}[O][O_3]. \tag{4.12}$$

These equations show that during the night [O_3] increases at the expense of [O] by virtue of reaction (4.2) because of the much greater O_2 concentration. However, since the supply of atomic oxygen is rapidly depleted, the increase in O_3 peters out (i.e. as [O] $\to$ 0, the right hand side of (4.12) $\to$ 0 and [O_3] saturates).

4.3 Ozone Depletion

4.3.1 *Thinning of the ozone layer and 'ozone holes'*

The effective thickness (i.e. in terms of absorption) of ozone is conventionally expressed in *Dobson Units* (DU), defined in terms of the corresponding thickness of ozone at standard temperature and pressure (STP). Specifically, the absorption of 100 DU of stratospheric ozone is equivalent to that of a 1 mm thick layer of ozone at STP. In Europe, the minimum safe level of ozone is defined to be 200 DU and concentrations below this are thought to be associated with an increased incidence of skin cancers. Ozone levels above Western Europe have been decreasing at an average rate of 0.8 DU per annum since the early 1970's and in the autumn of 1999, levels dropped briefly to 165 DU, well below what is considered to be safe.

Ozone production varies with latitude and season, primarily as a consequence of the distribution of solar irradiance. The O_2 photolysis rate will be considerably greater in equatorial regions where insolation is greatest and essentially unaffected by the seasons. The necessary reservoir of atomic oxygen will be correspondingly larger, and hence, so will the rate at which reaction (4.2) can proceed. Production at the poles will be substantially less and will be zero at the winter pole. We would therefore expect O_3 to diffuse away from equatorial regions towards the poles, (Fig. 4.5) where it should accumulate during the polar winter, equations (4.11) and (4.12). Polar O_3 levels at the end of winter should be well above 300 DU.

The Chapman reactions are based on a pure oxygen chemistry where only the different allotropes of oxygen are involved. Calculations of reaction rates suggest that the time constants for reaching equilibrium concentrations of O_3 vary with altitude, largely because the incident solar UV is attenuated during transmission through the stratosphere. Below about 30 km, the time constant is measured in weeks and a simple photochemical equilibrium is not possible. At these altitudes, the distribution of ozone is controlled by the dynamics of vertical and horizontal mixing. Above this altitude, the time constant becomes less than a day and chemical equilibrium should exist in spite of the diurnal variation in irradiance. However, measured concentrations of ozone are significantly less than the values calculated

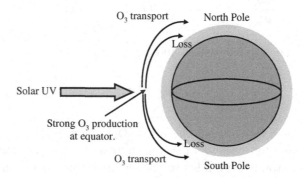

Fig. 4.5   Stratospheric circulation of ozone.

(i) Cl reacts with $O_3 \longrightarrow O_2 + Cl$

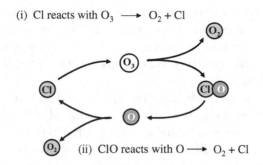

(ii) ClO reacts with $O \longrightarrow O_2 + Cl$

Fig. 4.6   Chlorine cycle for the catalytic destruction of ozone.

from (4.10) and the difference cannot be explained by uncertainties in the various rate constants and photolysis rates which are now well established.

4.3.2 *Chlorine cycle*

If the pure oxygen chemistry of the Chapman reactions, which is well understood, does not explain the low observed concentrations of stratospheric ozone concentration, then evidently other mechanisms for the breakdown of ozone must be involved. In 1974, Molina and Rowland, in a paper[6] for which they were later awarded the 1995 Nobel Prize, suggested that the photodissociation of chlorofluromethanes in the stratosphere released highly reactive free chlorine radicals that would catalytically destroy ozone.[a] Subsequently, the free radicals NO_x and HO_x were also shown to be important in the destruction of O_3. As *catalytic agents*, each of these free radicals is capable of destroying large numbers of O_3 molecules without itself being consumed. Of these, Cl is probably the most significant, because of its long lifetime in the stratosphere.

The Cl cycle is illustrated in Fig. 4.6. Free chlorine reacts strongly with ozone to produce chlorine monoxide (ClO) and an oxygen molecule;

$$Cl + O_3 = ClO + O_2 \qquad (4.13)$$

removing in the process an ozone molecule. The ClO molecule will in turn react with atomic oxygen to form another molecule of O_2, freeing the Cl atom for another reaction cycle.

$$ClO + O = Cl + O_2. \qquad (4.14)$$

Alternatively, two ClO molecules may react to release Cl

$$ClO + ClO = 2Cl + O_2. \qquad (4.15)$$

---

[a]The hypothesis had also been made independently by Stolarski and Cicerone.[7]

Thus, the complete cycle not only removes an O_3 molecule, it removes an oxygen atom as well, which is an essential and usually limiting component in the generation of O_3 (reaction 4.2). The residence time for a Cl atom in the stratosphere is in the order of 60–100 years, during which time it might catalyse the destruction of more than 10^5 O_3 molecules.

The principal sources of chlorine in the stratosphere are the halomethanes or halogenated hydrocarbons where one or more of the hydrogen atoms in methane have been substituted by halogens, e.g. CH_3Cl, CCl_4, and the chlorinated flurocarbons (CFCs) e.g. $CFCl_3$. The CFCs are particularly stable non-toxic compounds which have found widespread industrial applications. They are gaseous at ordinary temperatures and pressures but liquefy under modestly increased pressure, releasing latent heat in the process. When the pressure is released, they re-vapourise absorbing latent heat of evaporation. They are therefore ideally suited for use in heat pipes, refrigerators and air conditioners as the heat transfer fluid. Their stability has also made them suitable as the propellant gas in aerosol cans. In consequence, at the peak of production during the 1980's, millions of tons were manufactured each year, all of which would eventually be released into the troposphere. A small proportion would also diffuse across the tropopause and into the stratosphere.

Although highly stable in the troposphere, Rowland and Molina[6] showed that in the stratosphere the intense UV radiation will dissociate the molecules, releasing free Cl. For example,

$$CFCl_3 + h\nu = Cl + CFCl_2 \quad (h\nu > 4.3\,eV). \tag{4.16}$$

In due course, the remaining two Cl atoms will also be released by photolysis. Thus, the diffusion into the stratosphere of a single CFC molecule can introduce a number of free Cl atoms into the upper atmosphere, leading to the destruction of several hundred thousand ozone molecules.

Chlorine is removed from the stratosphere primarily by reaction with hydrogen to form HCl which then diffuses down into the troposphere where it is entrained in rainfall (contributing yet more environmental damage as acid rain!). Apart from elemental hydrogen, other sources of hydrogen available for the removal of Cl are methane (CH_4) and the hydrogenous radicals (HO_2) and (H_2O_2). In principle, fluorine is potentially just as destructive following analogous reactions to (4.13)–(4.15). In practice, the F reactions are not so critical because the stratospheric residence time is much shorter.

4.3.3 *Destruction of ozone by NO_x and HO_x reactions*

Of the other main ozone depleting processes, the NO_x cycle is one of the more important ones. The reaction is illustrated in Fig. 4.7. and described in reactions (4.17) and (4.18).

(i) NO reacts with $O_3 \longrightarrow NO_2 + O_2$

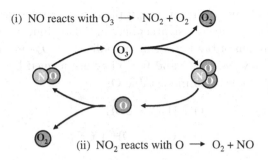

(ii) NO_2 reacts with $O \longrightarrow O_2 + NO$

Fig. 4.7 Catalytic destruction of ozone by NO_x.

$$NO + O_3 = NO_2 + O_2 \qquad (4.17)$$

and

$$NO_2 + O = NO + O_2. \qquad (4.18)$$

However, unlike the Cl cycle, photolysis of the NO_2 molecule can release O back into the stratosphere potentially generating another O_3 by (4.19).

$$NO_2 + h\nu = NO + O \quad (h\nu > 3.1\,\text{eV}) \qquad (4.19)$$

incidentally removing a photon of UV in the process. The principal loss mechanism for NO_x is by downward transport into the troposphere and by removal in rain as HNO_3.

The main natural source of NO_x is from nitrous oxide (N_2O) produced from biological activity at the surface of the Earth and the use of nitrogenous fertilisers. Microbial action converts organic nitrates (NO_3^-) mainly into N_2 but also some into N_2O. This diffuses into the upper atmosphere where it reacts with metastable oxygen produced by the photolysis of ozone (4.3) to produce two NO molecules.

$$N_2O + O = 2NO. \qquad (4.20)$$

The effects of this are mitigated by the competing (beneficial) photolysis reaction

$$N_2O + h\nu = N_2 + O \quad (h\nu > 5.4\,\text{eV}). \qquad (4.21)$$

The importance of this process as a source of stratospheric NO_x is unclear since it is critically dependent on transport into the stratosphere and unlike Cl, there are no highly stable transport agents such as the CFCs. Conversely, the intensive use of nitrogenous fertilisers means that the potential reservoir of N_2O is enormous.

The principal artificial source of stratospheric NO_x is probably from the exhausts of high flying jet aircraft. Typically, commercial jet airliners cruise at altitudes of 9–13 km close to the tropopause. Although most of the NO_x produced by jet aircraft probably remain in the troposphere where it acts as a greenhouse gas, a proportion will diffuse into the stratosphere and contribute to ozone depletion.

The single hydrogen of the hydrogenous radicals (HO_x) means they are highly reactive with 'odd' oxygen, i.e. O and O_3. They are formed by the dissociation of H_2O and CH_4 in reactions with metastable O:

$$O + H_2O = 2OH \tag{4.22}$$

$$O + CH_4 = OH + CH_3 \tag{4.23}$$

and once formed, catalytically destroy O and O_3 in a complex sequence of reactions

$$OH + O_3 = HO_2 + O_2 \tag{4.24}$$

$$HO_2 + O_3 = HO + 2O_2 \tag{4.25}$$

$$HO + O = H + O_2 \tag{4.26}$$

$$H + O_3 = HO + O_2 \tag{4.27}$$

$$H + O = HO. \tag{4.28}$$

The family of hydrogenous radicals is destroyed mainly by the pair of reactions

$$HO + HO = H_2O + O \tag{4.29}$$

$$HO + HO_2 = H_2O + O_2. \tag{4.30}$$

4.3.4 *The Antarctic ozone hole*

In the autumn of 1985, a team of British scientists working in Antarctica reported a reduction of about 50% in the stratospheric ozone layer over the South Pole; what has become known as the 'ozone hole'.[8] The hole was found to extend over most of Antarctica, an area comparable to the entire United States. Satellite mapping of stratospheric ozone[9] confirmed this. The hole continued to grow until by September 1996 it had increased to an area of $2.6 \times 10^7 \, km^2$ (an area greater than the whole of North America) with ozone levels as low as 111 DU in places. The ozone hole is now being seen as a major health hazard in Australia and regular ozone warnings are broadcast as part of normal weather forecasts. A similarly extensive hole has not been observed to form over the North Pole during the Arctic winter, although reductions in ozone levels in the order of 25% have been recorded.

There are special conditions that result in the production of the Antarctic ozone hole and explain why, despite some thinning of the ozone layer, a similar 'hole' has not formed over the North Pole. The hole is thought to be a consequence of the interaction between the chemistry of chlorine, nitrogen oxides and the Antarctic winter circulation.

During the southern hemisphere summer months, chemical reactions between chlorine monoxide and nitrogen dioxide trap chlorine into what has become known as chlorine reservoirs. Ozone depletion is as a result reduced during the late summer period. With the onset of the Antarctic winter, stratospheric circulation patterns change to form a vortex centred on the South Pole that inhibits latitudinal mixing. Residual gases in the stratosphere condense in the extreme winter cold to form the south polar stratospheric clouds composed of ice crystals. Despite the low temperatures, reactions take place on the surfaces of these crystals to liberate molecular chlorine (Cl_2) from the reservoirs.

At the beginning of spring, the clouds dissipate releasing the Cl_2 which is rapidly photolysed to atomic chlorine. The sudden increase in Cl levels rapidly depletes the ozone, creating the hole. By late spring/early summer, the winter vortex collapses drawing in ozone rich air from the mid-latitudes to restart the cycle. The South Pole ozone hole has steadily increased in extent and depth since it was first discovered, probably indicating a corresponding rise in the concentration of chlorine, in turn a consequence of the high levels of CFC use in the latter decades of the twentieth century.

A corresponding hole has not been observed to quite the same degree over the North pole probably because winter conditions are less severe. Arctic temperatures are not as cold and the winter vortex is less strong. Given the much higher population densities in the northern hemisphere, this is perhaps fortuitous.

4.4 Biological Impact of Ultraviolet Radiation

4.4.1 *Action spectra and damage*

The range of UV wavelengths absorbed by a particular molecular constituent of a biological cell is termed an *action spectrum* ($E(\lambda)$) and typically refers to some biological consequence, e.g. the action spectrum for sunburn, cataracts, skin cancer etc. Action spectra define the damage done to a biological system per unit intensity at a given wavelength.

In the environmental context, we are primarily concerned with the short wavelength part of the solar spectrum that overlaps with the relevant action spectra. If a given action spectrum lies outside the range of solar irradiation, then it is not a concern for life in general. What is at issue is the overlap between the action spectrum and the incident solar spectrum ($I(\lambda)$) known as the *solar effectiveness*, and given by the product $E(\lambda)I(\lambda)$ (Fig. 4.8).

We define the *Damage* (D) as the integral of the solar effectiveness.

$$D = \int_0^\infty E(\lambda)I(\lambda)d\lambda. \tag{4.31}$$

It is important to realise that the solar UV tail has a steep positive slope, while typical action spectra have strong negative slopes (in the wavelengths concerned) and thus a small reduction in the O_3 absorption of UV, which in effect moves the solar spectrum to shorter wavelengths, has a disproportionate impact on the

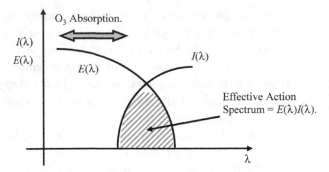

Fig. 4.8   Schematic diagram of action spectrum and solar UV tail, illustrating solar effectiveness.

damage. For example, a reduction in O_3 absorption by 10% leads to an increase of $\sim$45% in D for sunburn.

4.4.2 *Absorption by DNA and proteins*

A detailed description of the interaction between biological molecules and UV radiation lies outside the scope of this text and only a brief account will be given. The principal UV ($200\,\text{nm} < \lambda < 300\,\text{nm}$) absorption takes place on the four DNA bases, guanine (G), thymine (T), cytosine (C) and adenine (A) and peaks at a wavelength of $\sim$260 nm with a maximum extinction coefficient $\varepsilon(\lambda)$ (Sec. 3.2) in the order of $10^4\,\text{dm}^3\,\text{mol}^{-1}\text{cm}^{-1}$. The absorbed photon may result in oxidation of the base in question and potential mutagenesis. A major product is the pyrimidine dimer which impedes the correct reading and translation of the genetic code during cell reproduction, leading in some cases to the formation of cancerous cells, particularly, and most obviously, skin cancers.[10] Of the three types of skin cancer traceable to UV exposure, melanoma is particularly invasive and readily metastasises (forms away from the primary source) with a consequential high mortality. There has been a significant increase over the past quarter century in the incidence of skin cancers (both melanoma and other types) although it seems this is due only in part to ozone depletion. Changes in western lifestyles with holidays to sunny places and increased sunbathing have probably been a more significant factor.

About 10% of the absorption takes place in proteins such as the protein α-crystallin, an important constituent of the lens in the mammalian eye. Prolonged exposure will ultimately cause the formation of cataracts and impaired vision.

4.5 Ozone in the Troposphere

While the presence of ozone in the stratosphere is desirable as the UV shield, in the lower troposphere, it is a significant pollutant.[11] It is a strong oxidising agent that causes inflammation of the lung and linings of air passages, resulting in severe respiratory problems in high risk groups (the elderly, asthmatics, infants) and at higher concentrations, irritation of the eyes. It is also carcinogenic. Plant life and trees appear particularly susceptible to damage from air pollution and extensive forest and crop damage is frequently reported down-wind of large urban conurbations. Tropospheric ozone concentrations peak at altitudes of $\sim$1 km, suggesting that it is likely to be a major contributory factor in tree damage in mountain forests.

Ozone is one of the main constituents of smog, a photochemically induced fog formed by a complex chain reaction involving nitrogen oxides and water-based free radicals (HO_x, $x = 0$, 1, 2). The addition of NO_2 to almost any volatile organic compound (a major component in the exhaust of vehicles) will produce ozone when exposed to solar UV. Although there are both natural and anthropogenic sources of the precursor compounds, it is the latter arising from industrial and vehicle emissions that generates the high concentrations required to produce smog. As with most problems in pollution, it is the combination of a given pollutant with other environmental and medical factors that is usually the main issue.

4.6 Montreal Protocol

The potential threat from ozone depletion due to anthropogenic sources, principally Cl-containing transport agents, in the form of chlorinated fluorocarbons (CFCs) first began to be appreciated in the mid-1980's with the discovery of the Antarctic ozone hole. In response, the United Nations under its Environmental Programme convened a meeting in Montreal, Canada. The outcome was the Montreal Protocols[12] on the production, use and release of ozone depleting substances (ODS) such as CFCs. The protocol was signed and ratified by 140 countries, including the USA, making it the only universal agreement on the environment.

The original Montreal Protocols required the signatory countries to cut back production of ODS by 50% by 2000. A simultaneous programme of balloon and later satellite monitoring revealed, however, that this was insufficient and the conditions were tightened significantly in June 1990. This first amendment to the protocol required that developed countries should completely phase out ODS production by 2000 and in the case of developing countries, by 2010. Continuing concern led to a further amendment in November 1992 that brought forward the target date for phasing out production in developed countries from 2000 to 1996 and introduced timetables for the phasing out of other suspected ozone depleting agents. In particular, the hydochloroflurocarbons (HCFCs) that had been developed as supposedly less harmful replacements to CFCs. The HCFCs are now being replaced by chlorine-free compounds such as the hydroflurocarbons (HFCs). Concentrations of CFCs and other ozone depleting chemical appear to have peaked at around the year 2000 and it is projected that they will decrease steadily, albeit slowly, over the next several decades.

4.7 Problems

1 A particular type of sunscreen has an absorption coefficient α_s. Determine the thickness required to ensure a safe level of exposure, I_{Safe} for a given incident intensity of UV, I_0.

2 The attenuation of a beam of light by some absorbing medium is given by the
   Beer–Lambert Law which may be expressed as: $I = I_0 \exp(-\alpha)$, where I_0 and α
   are the incident (unattenuated) intensity and the absorption coefficient per unit
   thickness of the medium respectively.

   The intensity of incident UV light is measured at a particular ground weather
   station and found to increase approximately linearly over a period of t days.
   Assuming this to be due to thinning of the O_3 layer, derive an expression for
   the rate of change of α with time, $(d\alpha/dt)$.

3 If over a period of time Δt the change (assuming a linear change) in UV intensity
   is ΔI, show using the result of question 2 that:

$$\frac{\Delta I}{\Delta t} = \beta I(t = 0)$$

   where $1 + \beta = \exp(\alpha(0) - \alpha(\Delta t))$ and $a(t)$ is the absorption coefficient at time t.

References

1. http://rredc.nrel.gov/solar/spectra/1.5; NREL ASTM C173-03 Reference Spectra.
2. S. Chapman, A theory of upper-atmosphere ozone, *Mem Roy. Meteorol. Soc.* **3** (1930) 103.
3. S. Chapman, On ozone and atomic oxygen in the upper atmosphere, *Phil. Mag.* **10** (1930) 369.
4. S. Chapman, Some phenomena in the upper atmosphere, *Proc. Roy. Soc. A* **132** (1931) 353.
5. E. C. Y. Inn and Tanaka, *J. Opt. Soc Amer.* **43** (1953) 870.
6. M. J. Molina and F. S. Rowland, Stratospheric sink for chlorofluromethanes chlorine atom-catalysed destruction of ozone, *Nature* **249** (1974) 810.
7. R. S. Stolarski and R. J. Cicerone, Stratospheric chlorine: A possible sink for ozone, *Can. J. Chem.* **52** (1974) 1610.
8. J. C. Farman, B. G. Gardiner and J. D. Shanklin, Large losses of total ozone in Antarctic reveal seasonal ClO_x/NO_x interaction, *Nature* **315** (1985) 207.
9. R. S. Stolarski, A. J. Krueger, M. R. Schoeberl, R. D. McPeters, P. A. Newman and C. Alpert. Nimbus 7 satellite measurements of the springtime Antarctic ozone decrease, *Nature* **322** (1986) 808.
10. http://www.nas.nasa.gov/about/education/ozone/radiation.html
11. http://www.advisorybodies.doh.gov.uk/aom/index.ht
12. http//www.ozone.unep.org/pdfs/Montreal_Protocol2000.pdf

Chapter 5

HEAT TRANSFER PROCESSES

5.1 Modes of Heat Transfer

Heat energy always flows from places that are hot to places that are cold and the energy transfer takes place irrespective of whether the hot and cold regions are in physical contact. There are three fundamental mechanisms of heat transfer: radiative exchange, convection and thermal conduction.

Radiative processes are probably the most fundamental, since there will always be some exchange between a body and its environment so long as there is a difference in temperature between them. A surface at some temperature T will radiate energy according to the Stefan–Boltzmann Law which we first encountered in Chapter 3, equation (3.1):

$$J_R = \varepsilon \sigma T^4 \tag{5.1}$$

where J_R is the radiated current density (Wm^{-2}), σ the Stefan constant and ε is the emissivity. When $\varepsilon = 1$, the body is referred to as a black-body and rate of emission is a maximum. Real surfaces have $0 < \varepsilon < 1$ and are often referred to as 'grey-body' emitters. Two bodies in radiative contact will exchange radiant energy in accordance with (5.1) and if the two are at different temperatures there will be a net transfer of energy from the hotter to the colder:

$$J_R = \sigma(\varepsilon_h T_h^4 - \varepsilon_c T_c^4) \tag{5.2}$$

where the subscripts h and c refer to hot and cold respectively. Given the T^4 dependence, (5.2) reduces to (5.1) for all but the smallest differences in T. The small magnitude of σ (5.67×10^{-8} Wm^{-2}K^{-4}) means that at ordinary temperatures, radiative exchange is not large in relation to typical conductive and convective processes, although the latter do require physical contact.

Convective heat transfer processes rely on the flow of some fluid (liquid or gas) across a surface. Fluid particles in the boundary layer (i.e. the layer adjacent to the surface), will exchange kinetic energy with the surface resulting in heating or cooling of the surface and/or fluid depending on which is hotter. As the temperature of the fluid is changed, so will its density and hence buoyancy. The rate of energy transfer (J_{Cv}) is governed by Newton's Law of cooling:

$$J_{C_v} = h_c(T_h - T_c) \tag{5.3}$$

where h_c is the coefficient of convective heat transfer for the surface in question. Values of h_c are difficult to define, because in practice the rate of convection depends on local variables such as position, inclination etc. Equation (5.3) assumes convection is natural and not forced, by which we mean that any motion of the fluid is a direct consequence of the heat transfer and not as a result of some other external agency (e.g. a fan or pump). Convective heating is important in many ordinary situations such as cooking and domestic central heating systems based on radiators. Convection is important in nature as one of the principal heat transfer mechanisms within the troposphere. The formation of clouds, for instance, is largely a convective process (Sec. 2.3.3).

Convection requires the free displacement of atoms and molecules but since they are not free to move in solids, convection cannot occur. Instead, heat transport takes place by thermal conduction where kinetic energy is transferred between neighbouring atoms by lattice vibrations or by collision with mobile electrons. The more energetic the atoms and the electrons, the greater the collision and scattering rates become, and therefore, even though the individual events are random, there will be a net transfer of energy down the temperature gradient. This is expressed formally (for a homogenous medium) by Fourier's Law:

$$\underline{J}_\sigma = -k\nabla T \tag{5.4}$$

where $\underline{J}_\sigma$ is the conducted heat current density and k is the thermal conductivity of the medium $(\mathrm{Wm^{-1}K^{-1}})$. The minus sign indicates that heat flows down the temperature gradient.

5.2 Diffusion of Heat: Heat Equation

The rate of change of temperature at some position $(x,\ y,\ z)$ in a substance is governed by the conservation of heat, and hence that will be the result of the net effects of the three heat exchange mechanisms, and of any internal heat generation or heat sink processes, i.e.

Rate of change of heat content = net heat inflow + net internal heat generation

For simplicity, we will initially confine the discussion to the situation where there are no internal heat sources or sinks and assume that radiative and convective contributions can be neglected.

Consider the volume element (dV) in Fig. 5.1, the net heat inflow corresponds to the difference in $\underline{J}_\sigma$ flowing in from the left and that flowing out from the right of the element.

$$\underline{J}_\sigma dydz - \left(\underline{J}_\sigma + \frac{\partial \underline{J}_\sigma}{\partial x}dx\right)dydz = -\frac{\partial \underline{J}_\sigma}{\partial x}dV. \tag{5.5}$$

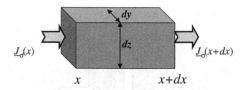

Fig. 5.1   Heat flow through a volume element.

The mass of the volume element is (ρdV) where ρ is the density and if the specific heat at constant pressure is c_p, then the energy required to raise the temperature of the element by one degree Kelvin is $(c_p \rho dV)$. Substituting for $\underline{J_\sigma}$ from (5.4) in (5.5), we get the one-dimensional rate of change of heat content in a homogenous medium (constant k)

$$\frac{\partial T}{\partial t} c_p (\rho dV) = -\frac{\partial \underline{J_\sigma}}{\partial x} dV = k \frac{\partial^2 T}{\partial x^2} dV. \tag{5.6}$$

Generally, there will be sources or sinks of heat within the volume element, $\dot{q}$ (Wm^{-3}) which must be included:

$$\frac{\partial T}{\partial t} c_p (\rho dV) = k \frac{\partial^2 T}{\partial t^2} dV + \dot{q} dV. \tag{5.7}$$

Dividing throughout by $(c_p \rho dV)$:

$$\frac{\partial T}{\partial t} = \frac{k}{c_p \rho} \frac{\partial^2 T}{\partial x^2} + \frac{1}{c_p \rho} \dot{q} = a \frac{\partial^2 T}{\partial x^2} + \frac{1}{c_p \rho} \dot{q} \tag{5.8}$$

where a is called the Fourier coefficient. Equation (5.8) is known as the one dimensional *heat equation*. In three dimensions:

$$\frac{\partial T}{\partial t} = a \nabla.(\nabla T) + \frac{1}{c_p \rho} \dot{q} = a \nabla^2 T + \frac{1}{c_p \rho} \dot{q} \tag{5.9}$$

where (∇^2) is the Laplace operator. Equations like (5.9) which relate the second derivative of space to the first derivative of time are widespread and apply to any diffusion process. We shall, for example, come across it again later in the text in relation to the diffusion of neutrons in a reactor, and of pollutants through the environment.

5.3 Examples of Heat Transfer by Conduction

5.3.1 *Double glazing*

It is unfortunate that the essential materials for house building, brick and concrete for strength and glass for light and windows, have high thermal conductivities. They are therefore poor insulators. One very effective solution is to use cavity wall

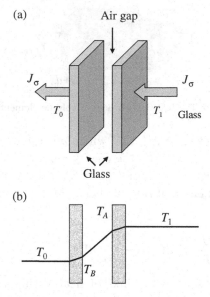

Fig. 5.2   Double glazed window: (a) heat flow (b) temperature profile.

insulation. The principle is to build two walls (adding to the strength) and fill the
space between them with some low thermal conductivity insulator.

The same idea is used in double glazing, except that the infill is air and the
units are generally evacuated and sealed to prevent convective exchange of air from
between the panes of glass and the outside. The situation is drawn schematically
in Fig. 5.2(a) for the common case of double glazing, where the inside and outside
temperatures are T_1 and T_0 respectively and T_A and T_B are the temperatures at
either side of the enclosed cavity. We assume that the heat current density, J_σ is
constant throughout the structure.[a] Using Fourier's Law in one dimension:

$$J_\sigma = -k_{glass}(T_1 - T_A) = -k_{air}(T_A - T_B) = -k_{glass}(T_B - T_0) \qquad (5.10)$$

where k_{glass} and k_{air} are the thermal conductivities of glass and air respectively.
Eliminating T_A and T_B from (5.10) we get for the overall heat current density:

$$J_\sigma = -\left(\frac{k_{glass}k_{air}}{2k_{air} + k_{glass}}\right)(T_1 - T_0). \qquad (5.11)$$

The thermal conductivity of glass is $\sim 1.4\,\mathrm{Wm^{-1}K^{-1}}$, while that for air is
$\sim 0.025\,\mathrm{Wm^{-1}K^{-1}}$, hence $k_{glass} \gg k_{air}$ and (5.11) reduces to

$$J_\sigma \approx -\left(\frac{k_{glass}k_{air}}{k_{glass}}\right)(T_1 - T_0) = -k_{air}(T_1 - T_0). \qquad (5.12)$$

---

[a]Formally the situation is identical to electrical current through a series connection of resistors.

In other words, heat flow is limited primarily by flow through the air cavity. It should be noted that this is only the case provided that the air cavity is properly sealed; failure to do so will allow comparatively warm air in the cavity to be continuously replaced with cold air from outside, effectively reducing the structure to a singly glazed unit.

The temperature profile through a double glazed window is shown in Fig. 5.2(b), which illustrates that most of the temperature difference is supported across the air gap.

5.3.2 *Periodic temperature variations: Annual cycle*

One very practical question which the heat equation allows us to address is how deep water pipes should be buried to ensure they do not freeze in winter. We treat the ground as a semi-infinite homogenous medium extending from the surface at $z = 0$ downwards to $z = \infty$, where we have taken the downward direction as positive.

The annual temperature cycle at the surface may be modelled, to a first approximation as a cosine variation about some mean value $\overline{T}$.

$$T(t) = \overline{T} + T_0 \cos(\omega t) \tag{5.13}$$

where ω, and T_0 are the angular frequency and amplitude respectively.

The solution to the one dimensional heat equation under these boundary conditions is a damped travelling wave:

$$T(z,t) = \overline{T} + T_0 \exp(-\alpha z) \cos(\omega t - k_\lambda z) \tag{5.14}$$

where k_λ is the wave vector. This is a heat wave that propagates into the ground with a velocity given by (ω / k_λ) and with an amplitude that diminishes as $\exp(-\alpha z)$. We note that the minimum temperature at the surface will be $(\overline{T} - T_0)$ and that as $z \to \infty$, the exponential tends to zero and the temperature deep underground will tend to the mean value, $\overline{T}$. In temperate latitudes this will be greater than freezing.

In order to determine how deep the pipe should be buried, we need to know the attenuation coefficient α and the propagation velocity of the wave. Substituting (5.14) into the heat equation (5.8) — with $\dot{q} = 0$ — and comparing the coefficients of the sine and cosine terms gives

$$v = \sqrt{2a\omega}; \quad \alpha = \sqrt{\omega/2a} \tag{5.15}$$

in terms of the known variables, ω (i.e. 2π ('year')$^{-1}$) and the Fourier coefficient for the soil (which in principle can be measured). For ordinary soils where $a \sim 1.4 \times 10^{-7}$ m^{2}s^{-1}, the wave propagation velocity is $\sim 2.4 \times 10^{-5}$ m s^{-1}; it would take some 49 days to penetrate a distance of 1 m into the ground. At this depth, the amplitude would have been reduced by over half, which in temperate climates implies that water pipes buried at a depth of a metre would probably not be at serious risk of freezing.

5.3.3 *Contact temperature*

A particularly interesting and important question is what happens when two surfaces at different temperature come into intimate contact. A common example would be when human skin comes into close contact with a hot (or cold) surface such as a hot mug or a chilled glass. The perception of temperature is primarily governed by the rate of heat transfer between the skin and the hot/cold surface. A metal object at room temperature will feel colder than a wooden one at the same temperature and ceramic floor tiles always seem colder to the feet than cork ones.

The situation may be considered as bringing into close contact two semi-infinite media, each with its own Fourier coefficient and initial surface temperature. Consider first a single semi-infinite medium stretching from $x = 0 \rightarrow \infty$ at a uniform initial temperature T_0.

Suppose at time $t = 0$, the surface temperature is abruptly changed to some new temperature T_1 and that after a sufficiently long time (i.e. t → ∞), all of the medium will attain the new temperature uniformly. Assuming there are no internal sources or sinks of heat ($\dot{q} = 0$),[b] then the solution of the heat equation takes the form:

$$T(x,t) = A + Bx + C \; \mathrm{erf}\left(\frac{x}{2\sqrt{at}}\right) \quad t \geq 0 \tag{5.16}$$

where A, B and C are constants to be determined from the boundary conditions.[c]

$$x = 0, \quad t \geq 0; \qquad T(0,t) = T_1$$
$$x > 0, \quad t = 0; \qquad T(x,0) = T_0$$
$$x > 0, \quad t \rightarrow \infty; \quad T(x,\infty) = T_1$$

The first boundary condition yields $A = T_1$, the second $C = (T_0 - T_1 - Bx)$ and the third, $B = 0$. Inserting these into (5.16)

$$T(x,t) = T_1 + (T_0 - T_1)\mathrm{erf}\left(\frac{x}{2\sqrt{at}}\right) \quad t \geq 0. \tag{5.17}$$

The rate of heat transfer into or out of the surface is given by Fourier's Law (5.4), i.e.

$$J_\sigma = -k\frac{\partial T}{\partial x} = -k\frac{\partial}{\partial x}\left\{(T_0 - T_1)\mathrm{erf}\left(\frac{x}{2\sqrt{at}}\right)\right\}. \tag{5.18}$$

The differential of an error function erf(β) with respect to its argument is:

$$\frac{d}{d\beta}(\mathrm{erf}(\beta)) = \frac{2}{\sqrt{\pi}}\exp(-\beta^2) \tag{5.19}$$

where in this instance, $\beta = x/2\sqrt{at}$.

---

[b]Clearly this would not be the case for human contact.
[c]Note erf(0) = 0 and erf(∞) = 1.

Using the result (5.19), we get the one dimensional heat flow:

$$J_\sigma = -k\frac{\partial T}{\partial x} = -k(T_0 - T_1)\frac{\partial \beta}{\partial x} \times \frac{\partial}{\partial \beta}(\text{erf}(\beta)); \tag{5.20}$$

$$J_\sigma = -(T_0 - T_1)\frac{k}{\sqrt{\pi a t}}\exp\left(-\frac{x^2}{4at}\right) = \frac{b}{\sqrt{\pi t}}(T_1 - T_0)\exp\left(-\frac{x^2}{4at}\right) \tag{5.21}$$

where $(b = \sqrt{k\rho c_p})$ is the *contact coefficient*.

When two bodies at temperatures T_1 and T_2 ($T_1 > T_2$) are brought into intimate contact, the interface will take some intermediate contact temperature T_C given by the condition that the heat flow out of the hotter surface just balances the heat flow into the cooler one ($J_{\sigma 1} = J_{\sigma 2}$). From (5.21), at the contact plane ($x = 0$)

$$\frac{b_1}{\sqrt{\pi t}}(T_1 - T_C) = \frac{b_2}{\sqrt{\pi t}}(T_C - T_2) \tag{5.22}$$

and the contact temperature is:

$$T_C = \frac{b_1 T_1 + b_2 T_2}{b_1 + b_2}. \tag{5.23}$$

Contact coefficients can be calculated for materials where the specific heat, density and thermal conductivity are known.

Equation (5.23) shows that where $T_1 \gg T_2$:

$$T_C \approx \frac{b_1}{b_1 + b_2}T_1 \tag{5.24}$$

which emphasises the need to use oven gloves (i.e. to ensure that $b_2 \gg b_1$) when removing hot pans from an oven!

5.4 Problems

1. The rear window of a car is made of laminated glass 5 mm thick with an effective thermal conductivity of $0.7\,\text{Wm}^{-1}\text{K}^{-1}$. A thin wire heater is attached to the surface of the window to demist/defrost it in cold or damp weather. Calculate the heat flow (J_σ) through the window if the temperature outside is $-1°\text{C}$ and the internal surface of the window is maintained at $6°\text{C}$. What power must be supplied to the heater if the rear window measures $1.4\,\text{m} \times 0.5\,\text{m}$, the free convection coefficient from the interior of the car to the window is $25\,\text{Wm}^{-2}\text{K}^{-1}$, and the temperature inside the car is $21°\text{C}$?

2. Derive the expressions (5.15) for the propagation velocity (v) and attenuation coefficient (α) assuming no heat sources or sinks in the ground.

3. The temperature at the surface of a floor is abruptly changed from T_0 to T_1 at time $t = 0$. Show that at some sufficiently long time t later, the rate of change of J_σ with x is approximately given by the expression:

$$\frac{\partial J_\sigma}{\partial x}\bigg|_t \approx \frac{b}{2a\sqrt{\pi}}(T_0 - T_1)\frac{x}{t\sqrt{t}}.$$

4. Two bodies with contact coefficients b_a and b_b at initial temperatures T_a and T_b are brought into contact. If $b_a T_a \gg b_b T_b$ and Ta is $1\frac{1}{2}$ times larger than the contact temperature, what is the ratio of the contact coefficients (b_b/b_a)?

Chapter 6

GENERATION OF POWER FROM FOSSIL FUELS

6.1 Introduction

In this chapter, the emphasis changes from an inherently passive approach where we try to understand our environment to a more proactive one, where we seek to develop ways and means, i.e. technologies to improve the quality of life. Energy is central to most modern technologies, whether in the home or the workplace. Most daily human activity depends on the availability of sufficient and reliable supplies of energy and apart from transport, this is generally electrical energy generated at some remote location and conveyed to the point of use through a highly integrated grid-based system.

This chapter will consider the generation of electrical power by large scale power stations burning hydrocarbon (i.e. fossil) fuels. Originally coal-fired, these stations are now frequently cleaner gas-fired systems, but whether in coal or gas burning, the generation processes share common aspects. Combustion of the fuel is used to heat a "working fluid", which drives a turbine connected to a generator producing the electricity. The fluid is then (in principle, at least) compressed/pumped and returned to the combustion chamber or boiler for the "cycle" to begin again. In gas turbine systems, the working fluid is the gaseous exhaust from the combustion process itself, while in coal fired stations superheated steam (steam at a temperature above the boiling point) is used instead. Although the latter utilise more complicated structures and are therefore more capital intensive to build, there are compensating advantages that reduce operating costs and make them cost effective options.

No matter how they are fuelled, power stations are governed by the Laws of Thermodynamics and we shall review these first. Following a brief introduction to the two principal laws and a number of their consequences, we shall consider some of the implications for efficiency, irreversible losses etc. and the impact these have on the performance of steam generating and gas turbine cycles.

6.2 Review of Thermodynamics

6.2.1 *The First Law of Thermodynamics*

The First and Second Laws of Thermodynamics govern the interchange of energy, work and heat between a system and its surroundings. We came across this earlier in deriving the lapse rate (Sec. 2.3.1), where we considered the vertical motion of a

71

parcel of air and the notion of stable and unstable atmospheres. In this section, we shall review some basic thermodynamics as it pertains to the operation of fossil fuelled power stations. These systems, large and complex as they are, may be modelled in terms of relatively simple thermodynamic cycles, where some working fluid is used to transfer energy from a hot source (boiler, combustion chamber) to a turbine which in turn drives a dynamo generating the electricity.

The First Law of Thermodynamics may be stated as:

When any closed system is taken through a cycle, the net work delivered to the surroundings is proportional to the heat taken from the surroundings.

For an infinitesimal amount of work delivered (δW) to the surroundings, we may write

$$\sum \delta Q \propto \sum \delta W. \tag{6.1}$$

The piston and cylinder illustrated in Fig. 6.1 is the archetypical example of a closed cycle system. Gas in the cylinder may be expanded by supplying some heat (δQ) from a heat source, pushing the piston outwards and so doing work (δW) on the surroundings. If the heat source is replaced by an appropriate heat sink, then the gas will cool transferring heat ($-\delta Q$) to the sink, and the surroundings will do work ($-\delta W$) on the piston pushing it back into the cylinder. The signs indicate whether work is being done by the system (positive) or being done by the surroundings (negative). The constant of proportionality in (6.1) is known as the 'mechanical equivalent of heat' and in S.I. units is just unity, i.e. both Q and W are measured in joules.

A corollary of the First Law of Thermodynamics (known as the first corollary) states that there exists a property of the system, the internal energy, U that is equal to the difference between the heat supplied to a system and the work done by the system on its surroundings:

$$\delta Q = \delta U + \delta W \tag{6.2}$$

i.e. the change in internal energy is the difference between the work done and the quantity of heat supplied. Note that U is a property of the closed (cylinder and piston) system, and is different in kind from heat and work. Internal energy is said to reside in the system and may be increased or decreased by a change of state.

The better known corollary of the First Law is the third which states that a perpetual motion machine of the first kind is impossible. Such a machine is one

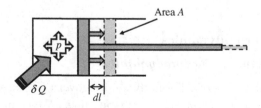

Fig. 6.1   Piston and cylinder illustration of the equivalence of energy and heat.

which, once set in motion, would continue to run for ever. It would, however, be of no practical value, since work could only be extracted from it at the expense of the internal energy and this would soon be exhausted in a few cycles.

Suppose some process changes the state of a system from state 1 to state 2 (e.g. expansion of the gas in the piston-cylinder to a new position), then (6.2) may be expressed in terms of the net quantities of heat and work crossing the boundary between the system and the surroundings

$$Q - W = U_2 - U_1 \tag{6.3}$$

where U_1 and U_2 are the internal energies of the initial and final states of the system. This is the non-flow energy equation as there is no net displacement of the fluid; in our example the gas expands and contracts but does not flow anywhere.

Where a system is in equilibrium, all the properties within it are uniform such that each unit mass has the same property, in particular, the same internal energy. The internal energy can then be written as $U = mu$ where m is the mass of the fluid and u is the specific internal energy (J/kg). It is generally convenient to work in specific quantities, i.e.

$$Q - W = u_2 - u_1 \tag{6.4}$$

where it is understood that Q and W refer to unit mass of fluid and u_1 and u_2 are the specific internal energies of the initial and final states. For reversible processes, (6.4) is more commonly expressed in differential form:

$$dQ - dW = du. \tag{6.5}$$

Let us return to the piston and cylinder illustrated in Fig. 6.1. Suppose we supply just sufficient heat to the gas to move the piston of area A out an infinitesimal amount dl with p assumed constant throughout the process. The force acting on the piston will be pA and the work done is $pAdl$, but since Adl is just the incremental change in the volume dV, then we get the more common form of the First Law of Thermodynamics

$$dQ = du + pdV. \tag{6.6}$$

Clearly, for a constant volume process where $dV = 0$, all the added heat will go into a change in the internal energy of the system.

$$dQ = du. \tag{6.7}$$

Conversely, in a constant pressure process, the added heat energy will be divided between a change in the internal energy and a change in the volume:

$$dQ = du + d(pv) = d(u + pv) \tag{6.8}$$

where v is the specific volume. The term $(u + pv)$ is the specific enthalpy, h and since it is defined as a combination of the properties u, p and v, it is by itself a property of the system. For a unit mass of fluid in a state of equilibrium undergoing a reversible process at constant pressure,

$$dQ = dh. \tag{6.9}$$

6.2.2 *The Second Law of Thermodynamics*

The Second Law of Thermodynamics expresses the fact that not all the available heat may be converted to work. The Law may be stated as:

It is impossible to construct a system which will operate in a cycle, extract heat from a hot reservoir and do an equivalent amount of work on the surroundings.

This implies that for a system to undergo a cycle and produce work, it must operate between at least two reservoirs of different temperatures, extracting heat from the hotter and rejecting some heat to the colder reservoir. A machine working continuously while exchanging heat with a single reservoir is known as a perpetual motion machine of the second kind and is not allowed.

A direct consequence of the Second Law is that there exists a property of a closed system, the entropy, S, defined as:

$$\int_1^2 \left(\frac{dQ}{T} \right) = S_2 - S_1 \text{ or in differential form,} \quad dS = \left(\frac{dQ}{T} \right). \tag{6.10}$$

Entropy defined in this way only applies to reversible cycles and represents the minimum entropy change. Real systems have irreversible losses, for example friction, and more generally (6.10) must be written:

$$dS \geq \frac{dQ}{T} \tag{6.11}$$

known as the Clausius inequality. Combining the First and Second Laws, we get for a reversible process

$$dQ = T ds = du + pdv \tag{6.12}$$

where s is the specific entropy. It is often useful to express (6.12) in terms of the specific enthalpy change, by substituting for the internal energy using (6.9)

$$T ds = dh - vdp \tag{6.13}$$

6.2.3 *Heat engine*

In a closed system passing through a number of thermodynamic states, the transfer of energy between successive states is governed by the First and Second Laws of Thermodynamics. Such a sequence of steps may be characterised as a series of operations in which heat Q_i received from a hot reservoir is used to perform some useful work W, and the unused (rejected) heat energy Q_o is passed to a cold

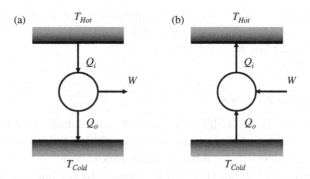

Fig. 6.2 (a) Heat engine; (b) Heat pump.

reservoir. Such a system operating in a closed cycle is referred to as a 'heat engine' and is illustrated in Fig. 6.2(a). It is possible to operate the system in the reverse direction, so that heat is transferred from the cold to the hot reservoirs by doing work as in a refrigerator (Fig. 6.2b).

Clearly, the larger the proportion of the received heat that is converted into work, the better or more efficient the engine. We define the thermal efficiency as the ratio of the work output over one cycle to the heat supplied over the same cycle

$$\eta = \frac{W}{Q_i} = \frac{Q_i - Q_o}{Q_i}. \tag{6.14}$$

Another way of expressing (6.11) is that the total entropy of a closed system must increase ($dS \geq 0$) and we can apply this principle to a heat engine system, extracting energy from a hot reservoir (temperature T_H), performing work and rejecting some heat to a cold reservoir (temperature T_C)

$$\frac{Q_o}{T_C} - \frac{Q_i}{T_H} = \frac{Q_i - W}{T_C} - \frac{Q_i}{T_H} \geq 0 \tag{6.15}$$

from which we can obtain the efficiency (6.14) in terms of the temperatures:

$$\eta = \frac{W}{Q_i} \leq \left(1 - \frac{T_C}{T_H} \right) \tag{6.16}$$

where it is clear that the maximum efficiency for a heat engine known as the Carnot efficiency occurs when

$$\eta_{\max} = \left(1 - \frac{T_C}{T_H} \right). \tag{6.17}$$

There are some important observations that may be made from the simple result (6.17):

Firstly, the average cold reservoir temperature is set by the atmosphere, i.e. T_C is fixed by nature. Secondly, to maximise the efficiency, one should minimise the second term, by ensuring $T_H \gg T_C$ and since the latter is fixed, then the heat engine should be operated at as high a temperature as possible. The efficiency of most heat engines is given by some relationship of the general form of (6.17) and thus should be run as hot as practicable.

6.2.4 *Vapour power cycles*

The common characteristic of all the power stations to be described in this chapter is that heat is transferred from a reservoir to a working fluid, which is then taken through a thermodynamic cycle during which work is done by the fluid, some heat is rejected to the cold sink and the fluid is returned to the hot reservoir. The cycle is illustrated in Fig. 6.3. The working fluid is first compressed/pumped ($1 \rightarrow 2$), requiring an input of work $-W_1$; acquires heat Q_i in the boiler or combustion chamber, ($2 \rightarrow 3$); is expanded through the turbine providing work out W_2 ($3 \rightarrow 4$) and then rejects heat Q_o to the sink (i.e. atmosphere) ($4 \rightarrow 1$) completing the cycle.

In a general way, we can represent such cycles in terms of pV diagrams, graphs of the variation in p with V from which the net work done around the cycle is the enclosed area (6.6).

$$W_2 - W_1 = \oint pdV = \int_1^2 pdV + \int_2^3 pdV + \int_3^4 pdV + \int_4^1 pdV. \qquad (6.18)$$

Alternatively, the cycle may be described in terms of the variation in entropy with temperature (6.12) round the cycle, the TS diagram, where the area enclosed is also equal to the work done.

$$W_2 - W_1 = \oint TdS = \int_1^2 Tds + \int_2^3 Tds + \int_3^4 Tds + \int_4^1 Tds. \qquad (6.19)$$

Steps $1 \rightarrow 2$ and $4 \rightarrow 1$ will be negative as they constitute the work required to compress/pump the fluid and the rejected heat component respectively.

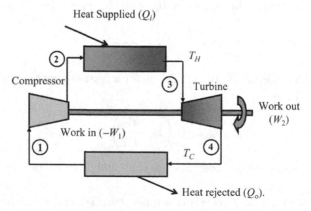

Fig. 6.3   Basic vapour power cycle.

Note that the vapour is no longer stationary, as in the simple example of the cylinder and piston of Fig. 6.1. The fluid is moving with some speed $\dot{x}$ and therefore, will have kinetic energy $(1/2m\dot{x}^2)$, in addition to the internal energy mu. The vapour may also have to move through a height z and so may gain or lose gravitational potential energy as well. Suppose that a small element of mass δm flows in with velocity $\dot{x}_1$ at an elevation z_1 and with enthalpy h_1. Simultaneously, another element of mass δm flows out with velocity $\dot{x}_2$ at elevation z_2 with enthalpy h_2, then

$$Q - W = (h_2 - h_1) + \frac{1}{2}(\dot{x}_2^2 - \dot{x}_1^2) + g(z_2 - z_1) \qquad (6.20)$$

where Q and W are the heat work transfers per unit mass flowing through the system. Equation (6.20) is the steady-flow energy equation. It is based on the following assumptions:

(a) Mass continuity: i.e. mass flow at inlet is equal to mass flow at outlet;
(b) The properties at any particular point within the system are time independent;
(c) Any heat and work crossing a boundary does so at a uniform rate;
(d) The properties across the cross section of the inlet and outlet are uniform.

In this context, a boundary is not just the physical barriers, for example, the walls of the boiler, but includes the 'open' parts, i.e. the inlet and outlet pipe cross sections. The properties of the fluid are only known at the open boundaries where they can be measured. Throughout the system, the mass flow rate is taken as constant (i.e. kg s^{-1}) even though particle speeds may vary dramatically. The properties at a given point in the system are therefore constant with time, although they will in general be different at different points in the system.

Under steady state operation assumptions (a), (b) and (c) hold reasonably well, although in some instances (b) should be interpreted as an average determined over a period of time that is long compared with any periodic fluctuations. However, when a fluid flows through a pipe, viscous effects cause the velocity to fall to zero in the layer adjacent to the pipe wall (known as the boundary layer) and as a result, the assumption (d) is not as robust as the other three.

In practical power systems work is extracted from the vapour as it is expanded from high, to low pressure in a turbine. If the fluid velocities are sufficiently high, there will be little time for the transfer of heat to the environment and the process is effectively adiabatic, i.e. $dQ = 0$. From the definition of entropy, (6.10) adiabatic processes are also isentropic (constant entropy) because for reversible processes $dS \propto dQ$. High speed turbines operate by directing high speed jets of working fluid onto a set of curved blades attached to a rotor. The jets of fluid change their direction, exerting a force on the blades and imparting a torque to the shaft as a result, (Fig. 6.4a). High speed jets can be generated by passing fluid through ducts of decreasing cross sectional area. Mass continuity (assumption (a)) requires that the speed must increase, and if the system is essentially adiabatic, then there must be a corresponding reduction in the pressure (Fig. 6.4b): the increase in the kinetic energy must be balanced by a corresponding reduction in the enthalpy.

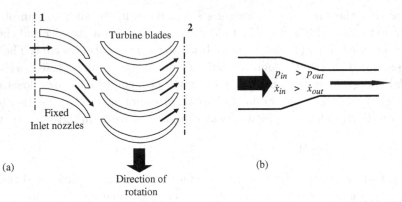

Fig. 6.4 Principle of isentropic, high speed turbine: (a) fixed nozzles direct high speed jets onto turbine blades; (b) reduction in duct cross section induces pressure drop and accelerates the flow.

The reduction in the fluid momentum means that the velocity on exit from the turbine is similar to that it possessed just before entering the nozzles, i.e. $\dot{x}_1 \sim \dot{x}_2$. The associated loss of kinetic energy is just that transferred to the rotational energy of the turbine (since $dQ \sim 0$). The steady flow equation (6.20) reduces to:

$$W_2 \approx (h_1 - h_2). \tag{6.21}$$

The work produced is equal to the reduction in enthalpy of the vapour as it passes through the turbine. Compressors are essentially taken as the reverse process, where the added velocity imparted to the fluid by the rotating vanes is reduced by diffusers (reversed nozzles) and the pressure increased correspondingly. The process is still effectively isentropic, and described by (6.21), except that W is negative because work has to be done on the fluid.

6.2.5 *Carnot cycle*

In principle, the Carnot cycle illustrated in Fig. 6.5(a) is the most efficient vapour-power cycle. The fluid is compressed isentropically, consuming some work (W_{12}), heated (Q_{23}) isothermally (constant temperature), expanded isentropically through a turbine to produce the work (W_{34}) and then heat is rejected to the environment again isothermally (Q_{41}). Thermodynamically, the cycle is depicted in Fig. 6.5(b).

Adding heat isothermally to a system may be achieved by generating steam in a boiler. The temperature will remain constant as all the added heat will be used in boiling the water. The boiling point temperature increases with pressure, hence the need for a compressor to raise the pressure and in turn, the temperature and thermodynamic efficiency (6.17).

The net work produced and the heat input in the Carnot cycle may be calculated from the TS diagram, Fig. 6.5(b).

$$W_{34} - W_{12} = T_H(S_3 - S_2) - T_C(S_4 - S_1); \quad Q_{23} = T_H(S_3 - S_2) \tag{6.22}$$

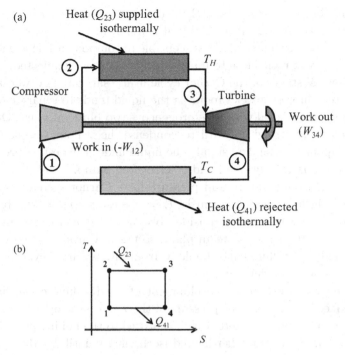

Fig. 6.5   Carnot cycle: (a) schematic diagram of cycle; (b) T-S diagram.

and the thermodynamic efficiency is

$$\eta = \frac{W_{34} - W_{12}}{Q_{23}} = \frac{T_H - T_C}{T_H} = 1 - \frac{T_C}{T_H} \tag{6.23}$$

this is just the Carnot efficiency (6.17). It is important to appreciate that the maximum efficiency is achieved in the Carnot cycle, because heat is transferred to and from the working fluid under isothermal conditions. If the temperature is not maintained during these parts of the cycle, then the efficiency will always be lower.

6.3 Steam Generating Power Stations: Rankine Cycle

The largest, and probably still the most common of the fossil fuel fired stations are those which burn coal to produce steam, which in turn is used to drive the turbines and hence the generators that produce the electrical power. The Carnot cycle is not used in these systems for two principal reasons. It has a low work ratio and there are difficulties with the compression stage of the cycle.

Work ratio is the ratio of the net work $(W_{34}-W_{12})$ to the positive work produced by the turbine, W_{34}. The thermodynamics reviewed in the preceding sections applies strictly only to reversible (ideal) systems, where the system may be returned to its initial state in all aspects by reversing the cycle. Real systems have irreversible losses due to, for example, friction in mechanical components. It follows that processes with low work ratios are particularly susceptible to component inefficiency, whereas a system with a work ratio close to unity will be relatively unaffected.

The isothermal stages in the Carnot cycle utilise saturated or 'wet' steam. It is very difficult to compress a wet vapour as the liquid tends to separate out and the compressor has then to deal with a heterogeneous two phase mixture. On the other hand, it is relatively straightforward to condense the vapour completely, remove any residual vapour fraction and supply the liquid under the desired pressure to the boiler using a comparatively small and efficient feed pump.

The maximum temperature and pressure for a Carnot system are set by the critical point, which for water/steam is $T_c = 647K$ and $P_c = 221$ bar. Modern materials can operate reliably at considerably higher temperatures and pressures ($T \sim 900K$, $P \sim 350$ bar) for a steam plant, so that a Carnot system would under utilise the available metallurgical technology. Ironically, a Carnot cycle power station would not be using fuel efficiently.

These deficiencies led to the development of the Rankine cycle illustrated in Fig. 6.6. The feed pump raises the pressure of the water isentropically ($1 \rightarrow 2$; W_{12}) and supplies it to the boiler, where it is first heated to the boiling point isobarically ($2 \rightarrow 3$; Q_{23}). The steam is further heated isothermally until dry ($3 \rightarrow 4$; Q_{34}) and finally superheated isobarically to raise the temperature to the desired operating value ($4 \rightarrow 5$; Q_{45}). The superheated steam is expanded isentropically through the turbine to provide useful work ($5 \rightarrow 6$; W_{56}) and generate electrical power. The

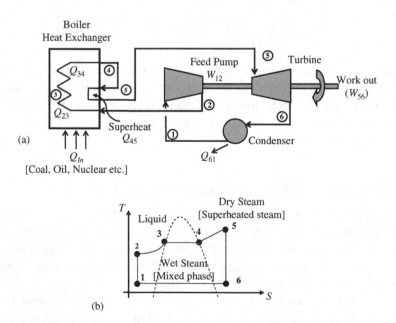

Fig. 6.6   Rankine cycle: (a) schematic illustration; (b) $T - S$ diagram.

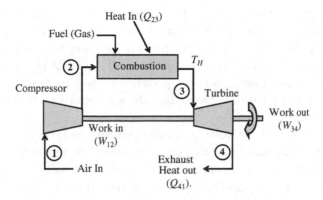

Fig. 6.7   Joule (Brayton) gas turbine cycle.

steam emerges from the turbine outlet both cooled and wet; the remaining moisture is condensed out isothermally in the characteristically shaped cooling towers and returned to the feed pump ($6 \rightarrow 1$; Q_{61}) to start the cycle again.

One of the major advantages of the Rankine cycle is that the feed pump is pressurising a liquid and hence there is no change in the volume of the working fluid, $\delta V = 0$. The feed pump can be relatively small and the work input ($-W_{12}$) is equally small compared with the turbine output (W_{34}), leading to a high work ratio.

In addition, superheating allows much better utilisation of the combustion chamber and improves the efficiency dramatically. Although there are additional irreversibilities associated with superheating, since the work ratio in the unsuperheated Rankine cycle is so high, the small additional demand makes little significant difference. More importantly, the net work per unit mass of steam is much greater, so that the added complexity and cost of installing the superheater is adequately compensated by a reduction in the size of other components.

6.4 Gas Turbine Generating Systems: Joule Cycle

Unlike steam generating power systems, gas turbine stations utilise the hot exhaust gases from the combustion stage to drive the turbine. Temperatures are much higher and superficially one might suppose that efficiencies should be greater, although in practice this is not the case.

In some aspects a gas cycle more closely resembles a vapour one than does the Rankine; the working fluid is compressed, heated through the combustion of an air-gas mixture, expanded through the turbine and cooled to the low temperature reservoir.

The basic principle of a gas turbine cycle, known as the Joule (Brayton) cycle, is illustrated in Fig. 6.7. The working fluid is air, which is pressurised ($1 \rightarrow 2$) consuming work W_{12}, mixed with the fuel and burnt in the combustion chamber under isobaric conditions ($2 \rightarrow 3$), adding heat Q_{23}. The hot exhaust gas is then expanded through the turbine ($3 \rightarrow 4$), generating work W_{34} and simultaneously undergoing cooling. Finally, the exhaust gases are released into the atmosphere ($4 \rightarrow 1$), carrying the rejected heat, Q_{41}. Strictly speaking, the cycle is not a closed one, but so long as the added mass of fuel is small (which it is) in relation to the mass of air, and the air pressure and temperature at the inlet to the compressor are the same as that of the cold reservoir, then the approximation to a closed cycle is acceptable.

The use of an internal combustion chamber, where heat is transferred by burning the fuel in the working fluid, coupled with the open cycle where the exhaust gases are released directly into the atmosphere, mean that the bulky boiler and cooling stages employed in the Rankine cycle can be dispensed with. These characteristics make gas turbine systems somewhat simpler and less expensive, making them ideally suited for peak load operation, where capital cost is more important than operating costs.

The work ratio in gas turbine systems is, however, much lower than in the Rankine system, largely because the working fluid is a compressible gas rather than an incompressible fluid. An appreciable proportion of the work output from the turbine is therefore consumed in the compression stage of the cycle. In addition, heat is not transferred to the working fluid at constant temperature and hence the cycle efficiency will be somewhat lower than the Carnot cycle efficiency, which simply depends on the temperature ratio (Sec. 6.2.5).

It is more straightforward to estimate the efficiency in gas turbine cycles since effectively the working fluid (i.e. air or nitrogen) does not undergo any phase change (i.e. as in water/steam). Consequently, the specific heats at constant pressure change relatively little around the cycle. In the Rankine cycle, calculations of efficiency must be derived from previously determined steam tables and cannot be easily calculated analytically. The efficiency of a Joule cycle, can be estimated from the TS diagram, Fig. 6.8(a):

$$\eta = \frac{W_{34} - W_{12}}{Q_{23}} = \frac{(h_3 - h_4) - (h_2 - h_1)}{Q_{23}}$$
$$= \frac{c_p(T_3 - T_4) - c_p(T_2 - T_1)}{c_p(T_3 - T_2)} = \frac{(T_3 - T_4) - (T_2 - T_1)}{(T_3 - T_2)}. \tag{6.24}$$

where c_p is the specific heat at constant pressure and we have assumed that the turbine and compressor processes are isentropic.

The pressure ratio $r_p = (p_2/p_1) = (p_3/p_4)$ (Fig. 6.8(b)) for isentropic compression and expansion may be related to the temperatures using the adiabatic relationship $T^\gamma p^{(1-\gamma)}$ is a constant[a] to give

$$T_2 = T_1 r_p^{(\gamma-1)/\gamma} \quad \text{and} \quad T_3 = T_4 r_p^{(\gamma-1)/\gamma} \tag{6.25}$$

---

[a]For an adiabatic process, pV^γ = constant. From the Ideal Gas Law, $V = RT/p$, substitution for V gives $T^\gamma p^{(1-\gamma)}$ = constant

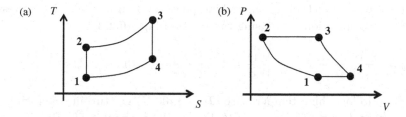

Fig. 6.8   Joule (Brayton) gas cycle : (a) $T - S$ and (b) $p - V$ diagrams.

where γ is the ratio of the specific heat at constant pressure to the specific heat at constant volume. Rearranging (6.24) and substituting for T_2 and T_4 in (6.24) we get, for the ideal efficiency

$$\eta = \frac{(T_3 - T_2) - r_p^{(1-\gamma)/\gamma}(T_3 - T_2)}{(T_3 - T_2)} = 1 - r_p^{(1-\gamma)/\gamma}. \tag{6.26}$$

Clearly, the compression ratio is an important parameter in determining the efficiency of a gas turbine system and it is important to know what constitutes the optimum value. Note that this is not simply the maximum value of $r_p = (T_3/T_4)$ as in that case all the turbine output would be consumed by the compressor!

$$W_{34} - W_{21} = c_p(T_3 - T_1) - c_p(T_2 - T_1) = c_p r_p^{\gamma/(1-\gamma)}(T_4 - T_1) = 0. \tag{6.27}$$

The net work produced ($W_0 = W_{34} - W_{21}$) from a gas turbine system is:

$$\begin{aligned} W_0 &= c_p(T_3 - T_4) - c_p(T_2 - T_1) \\ &= c_p T_3(1 - r_p^{(1-\gamma)/\gamma}) - c_p T_1(1 - r_p^{(\gamma-1)/\gamma}). \end{aligned} \tag{6.28}$$

Differentiating and equating to zero in the usual way to find the maximum:

$$\frac{dW_0}{dr_p} = c_p T_3 \frac{\gamma - 1}{\gamma} r_p^{(1-2\gamma)/\gamma} - c_p T_1 \frac{\gamma - 1}{\gamma} r_p^{(-1/\gamma)} = 0. \tag{6.29}$$

From which the optimum value of r_p is

$$r_{p_{\text{opt}}} = \left(\frac{T_3}{T_1}\right)^{\gamma/2(\gamma-1)}. \tag{6.30}$$

The temperature T_3 may be as high as 1400 K in a modern gas fired turbine power system which, taking the ambient temperature T_1 as 283 K, gives a theoretical optimum compression ratio for maximum work output of ~ 16.4. The corresponding ideal efficiency would be around 55%. This is never achieved in practice because of irreversible losses in the system, for example because turbines and compressors are not truly isentropic devices. The sensitivity to irreversible losses is represented by the work ratio, r_w which for a gas turbine power system may also be expressed in terms of the compression ratio using the relationships (6.25)

$$r_w = \frac{W_{34} - W_{12}}{W_{34}} = 1 - \frac{W_{12}}{W_{34}} = 1 - \frac{(T_2 - T_1)}{(T_3 - T_4)} = 1 - \frac{T_1}{T_3} r_p^{(\gamma-1)/\gamma}. \qquad (6.31)$$

The work ratio for a high temperature ($T_3 = 1400$ K) gas turbine operating at the optimum compression ratio ($r_p \sim 16.4$) would be about 0.55. By comparison, the work ratio for a well optimised Rankine cycle might be well in excess of 0.8. Consequently, actual efficiencies for gas turbines are rather less than the theoretical values and in fact comparable to steam generating systems.

It is the case that in principle, the higher the temperature at the turbine inlet valve (T_3), the better will be the performance of a gas turbine system. The work ratio, optimum compression ratio and efficiency are all increased by operating at a higher turbine inlet temperature. The upper temperature limit is determined mainly by the metallurgical properties of the materials from which the turbine components are made. Inevitably there is a trade-off between performance and cost, since in general higher performance materials are more expensive, require higher levels of maintenance and are more short-lived. Although materials costs may limit the operating temperatures and potential efficiencies, gas turbine systems are comparatively inexpensive to build in relation to more complex steam generating power stations, but operating costs are greater. Gas turbine systems, on the other hand, may be switched on and off rapidly, making them well suited to peak load use.

6.5 Combined Power Cycle Generating Stations

The exhaust temperature at the turbine outlet (T_4) of a gas powered station is still high and may be usefully employed as the source of heat in a secondary steam generating system, considerably increasing the overall fuel efficiency. Typically, the turbine exhaust gas temperature might be ~ 900 K, sufficiently hot to generate steam to drive a steam turbine system. In fact, it is often advantageous in such systems to provide additional heat to the steam generating boiler in order to raise the overall thermal efficiency to the best achievable.

The basic arrangement is illustrated in the block diagram of Fig. 6.9. Combustion of gas supplies heat Q_{Gas} to the gas turbine system, yielding useful

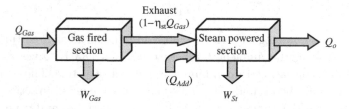

Fig. 6.9   Block diagram of combined power cycle generating systems.

work $W_{Gas} = \eta_{Gas}Q_{Gas}$ where η_{Gas} is the efficiency of the gas cycle. The balance of the input energy $(1 - \eta_{Gas})Q_{Gas}$ is in principle available to generate steam in the second stage, although in practice some of this will be lost. If some additional heat Q_{Add} is supplied to the steam boiler, then the total energy input to the steam generating second stage will be $(Q_{Add} + \eta_{Tr}(1 - \eta_{Gas})Q_{Gas})$ where η_{Tr} represents the efficiency of heat transfer between the gas and steam sections. Taking $\eta_{Tr} = 1$ as a limiting case, the overall efficiency is

$$\eta_T = \frac{W_{Gas} + W_{St}}{Q_{Gas} + Q_{Add}} \approx \frac{\eta_{St}(Q_{Gas} + Q_{Add}) + \eta_{Gas}Q_{Gas}(1 - \eta_{St})}{Q_{Gas} + Q_{Add}}$$
$$= \eta_{St} + \frac{\eta_{Gas}Q_{Gas}(1 - \eta_{St})}{Q_{Gas} + Q_{Add}} \tag{6.32}$$

where η_{St} is the efficiency of the steam plant. If the individual stage efficiencies are typically 35%–40%, then an overall efficiency of $\sim 55\%$ can be achieved representing a significant improvement. However, dual cycle generating stations are considerably more complicated and correspondingly expensive to construct. The fact that the steam generating component of the system cannot readily be switched on or off, the high construction costs and high fuel efficiency make these stations well suited for use as base or intermediate base load supply.

6.6 Distribution of Electrical Power

The large scale distribution of electrical power is invariably based on a grid system where power is generated at intersecting nodes by a comparatively small number of sizeable central power stations. In general, these will be a mix of thermal (i.e. fossil fuel fired), nuclear, and where terrain and rainfall permit, hydroelectric facilities.

Grid distribution systems are inherently resistant to failures at source. Should a power station on the grid 'go down' for some reason, the power deficit can be supplied from elsewhere on the system where generating capacity is presently under utilised. Grid systems are also flexible, enabling sudden surges in demand to be serviced rapidly and safely, and allow for planned removal of generating capacity in order to carry out routine maintenance.

Management of the grid supply is a crucial aspect of the electrical supply system. Sufficient capacity must be maintained to meet fluctuations in demand and provide for unexpected contingencies, without undue levels of costly redundancy within the system. This is commonly achieved through a combination of base load and peak load generation. Base load supply is provided by large facilities, typically nuclear and large coal fired systems that are not well suited to rapid changes in output. On the other hand, these stations can be configured to run steadily at high levels of efficiency, and thus with good fuel utilisation. Fluctuations in demand are supplied mainly by gas turbine and hydroelectric stations that can be switched on to the grid quickly, without damage to the infrastructure. Properly managed, a grid system can provide an optimal mix of base and peak load provision with a consequent optimal use of fuel, while simultaneously guaranteeing continuity of supply on demand to the user.

Grid systems offer one other significant advantage. Electricity is difficult to store. Batteries are an extremely inefficient means of storage. They are expensive, bulky (with low energy densities) and have comparatively short working life. They are suitable for low power applications where portability is the prime requirement, but they are completely unsuited for high power storage applications and are therefore not used in these instances. Pump-release hydroelectric systems, in contrast, are quite widely used to store excess generated power. In these systems, spare electrical capacity is used to pump water up into the dam (i.e. converting electrical energy into gravitational potential energy). The water is then released through turbines to produce power when there is a shortfall in supply. Pump-release systems can respond very rapidly to changes on the grid, the turbines often being allowed to 'freewheel' when not generating power, to minimise acceleration times. Pump-release systems are typically used to store excess base load production from power systems that are not easily switched off.

6.7 Sequestration of Carbon Dioxide

Any hydrocarbon burning power station produces carbon dioxide, and its removal is a subject of major concern. Technologies do exist and are being developed to eradicate or *sequester* CO_2 from the flue gases produced by a power station. There are two stages in this process. Firstly, the CO_2 has to be separated or *captured* from other gases in the exhaust, mainly water vapour and nitrogen; the other major components as combustion will have taken place in air. Typically 10–12% of the exhaust from a coal fired station will be CO_2 and somewhat less, 3–6%, in a gas fired system. Carbon capture is usually achieved using amine absorbers and cryogenics. It is expensive and accounts for $\sim 75\%$ of the overall cost of the sequestration procedure.

Secondly, the captured CO_2 has to be stored, usually by pumping it into deep geological formations such as oil and gas reservoirs, unrecoverable coal seams and deep saline reservoirs. Injecting CO_2 into an oil or gas reservoir will force the oil to the surface, an approach known as *enhanced oil recovery* and it has the advantage of offsetting some of the costs of sequestration. The carbon dioxide will remain in the reservoir provided the pressure does not exceed the original pressure. Currently, about 3×10^{11} kg of USA oil is extracted by some form of enhanced oil recovery. Carbon dioxide can also be used to extract methane adsorbed onto the surface of coal in seams that for various reasons cannot be mined (e.g. too thin). Carbon dioxide has an adsorption rate onto coal about twice that of CH_4 and so will displace the latter. The CO_2 remains fixed onto the coal, displacing the CH_4 which may be extracted and used either as a feedstock for the production of hydrogen or directly as a fuel. Although yet to be fully developed, this technology would also mitigate some of the storage costs.

The injection of CO_2 into deep saline reservoirs does not have any cost advantages but the potential storage capacity is believed to be substantial. Estimates of capacity in the USA alone are thought to amount to $\sim 5 \times 10^{13}$ kg. In Europe, the Norwegian company Statoil has been using the method since 1996 to inject $\sim 9 \times 10^8$ kg per annum of CO_2 into the Utsira Sand sea floor saline aquifer near its Sleipner West Heimdel gas reservoir (this corresponds to the output from a ~ 150 MW power station).

Other more controversial methods of sequestration are being researched as well, including the direct injection of CO_2 into deep oceans. It is generally believed that 80–90% of the atmospheric CO_2 will be ultimately dissolved into the ocean surface, eventually being drawn down into the deep ocean. However, the kinetics of this process are very slow and it is estimated that atmospheric levels would peak in several hundred years time. The idea is that direct injection would speed up the process. Many environmentalists fear this method would yield disastrous unintentional consequences. At the point of injection, the CO_2 would probably cause depressed levels of pH (i.e. increased acidity) seriously affecting the local chemical and biological equilibrium with the possibility of eutrophia, toxic blooms etc. It is difficult to imagine this particular method proving to be an acceptable or viable option.

The technology of carbon sequestration is still relatively new and unproven but the use of carbon sequestration as a means of extracting oil from otherwise difficult oil fields is a particularly attractive option, since it provides a cost effective way of achieving two desirable aims at little extra expense. It is obviously imperative that carbon dioxide injected into a deep geological feature be stable and much research still needs to be undertaken to confirm that this is indeed so.

6.8 Summary and the Future of Fossil Fuel Burning Power Stations

In this chapter, we have reviewed the two fundamental laws of thermodynamics, introducing important ideas such as the equivalence of heat and work (6.1), internal energy (6.2), enthalpy (6.9) and entropy (6.10). These ideas are incorporated in the notion of a heat engine, an abstraction that represents all thermodynamic systems capable of delivering useful work. Analysis of the ideal heat engine performance showed that there is an absolute maximum theoretical efficiency, the Carnot efficiency (6.16). The ideal heat engine is founded on the principle of a reversible process, where the working fluid is taken through a succession of thermodynamic states and returned precisely to its initial state. Real systems are not ideal but subject to irreversible losses, this is expressed in the Clausius inequality (6.11).

In the context of power stations, the working fluid is circulated between the source of heat, and the turbine driving the electrical generator, transferring enthalpy from one to the other. The fluid is moving round a circuit, and therefore will have kinetic energy, which when taken into account (together with any changes in gravitational potential energy), gives the equation of steady flow (6.20). Implicit in this equation is the concept of a boundary, across which heat flow will be uniform and time independent. Underpinning both of these is mass continuity; that the rate at which mass flows past any point in the system must be constant. It is why nozzles can be used to accelerate the flow at the inlet to the turbine, transforming enthalpy into kinetic energy, and so minimising the transfer of heat from the working fluid to the turbine.

In principle, there are two classes of power stations: those where the working fluid (usually steam) is heated externally (the Rankine cycle), and those where the heat is supplied by internal combustion (Joule, Brayton cycle). In the former, the source of heat is in a sense immaterial, usually it is coal or nuclear (Chapter 7), although waste incineration is not infrequently used and where geological conditions permit, it could be geothermal (Sec. 8.8). Systems based on the Rankine cycle, though expensive, are well understood and when properly designed, have high work ratios and good fuel utilisation.

Power systems based on the internal combustion of air/gas mixtures are potentially more efficient, but in practice suffer poor work ratios limiting the fuel utilisation. On the other hand, these systems are less complicated and relatively quick and inexpensive to build. The overall cost is critically dependent on the price of gas, and while this remains low, these systems are more competitive — hence the 'dash for gas', especially when compared with the high capital investment and consequential risk associated with a coal-fired station. The best use of fuel is achieved in the combined cycle power stations where the still hot exhaust from a gas fired station is used (often augmented) as a source of heat in a steam generating plant. Such combined cycle stations have particularly high efficiencies, but are expensive and even more complex structures to construct.

The efficiency of power stations has steadily increased as improved materials and newer designs provide better fuel utilisation. Incremental advances may be expected in the future. However, we have seen in this chapter that the laws of thermodynamics place upper limits on what may be achieved. The Carnot efficiency is the theoretical maximum that cannot be bettered, but unfortunately the Carnot cycle *per se* is not practical. Real systems must allow for irreversible losses, represented by the work ratio, and as is usually the case, the most optimum designs involve a trade-off between conflicting requirements.

There are two overarching issues surrounding the use of hydrocarbon fuels in general, whether for the generation of power or for transportation (Chapter 9). Firstly, there is the obvious question of the emission of carbon dioxide and the impact it will have on the global climate. As just discussed, sequestration technologies are being developed which offer at least the possibility of controlling the CO_2 released from power stations. The technologies may in the event prove ineffective, and will inevitably be costly. The second major concern is the steady depletion of fossil fuel supplies. Oil and gas supplies are still substantial but they are finite, and while consumption is increasing, they cannot be replenished on any timescales other than the geological. We shall be returning to this question in the final chapter.

6.9 Problems

1. (a) Show that for an ideal gas at constant temperature T:

$$\left(\frac{\partial p}{\partial V}\right)_T = -\frac{p}{V}$$

   where p and V are the pressure and volume respectively.

   (b) For any substance undergoing a reversible adiabatic (isentropic) process:

$$c_V \left(\frac{\partial p}{\partial V}\right)_S = c_p \left(\frac{\partial p}{\partial V}\right)_T$$

   where c_V and c_p are the specific heats at constant volume and constant pressure. Use this together with the result from part (a) to derive the Poisson expression:

$$pV^\gamma = K$$

   where $\gamma = c_p/c_v$ and K is a constant.

2. A steam generating power station delivers superheated steam to a single stage turbine at a pressure of 50 bar and a temperature of 450°C for which the tabulated specific enthalpy is $3316\,\text{kJ kg}^{-1}$. This falls to $2109\,\text{kJ kg}^{-1}$ after the turbine stage and a further $1988\,\text{kJ Kg}^{-1}$ is lost in the condenser, the water from which, is supplied to the feeder pump at a pressure of 0.04 bar. Assuming that the steam cycle can be described by an ideal Rankine cycle, estimate the net work done per cycle and the efficiency. (Calculate all values in a per kg basis.)

3. In a combined cycle gas turbine/steam generating power station, the steam is generated from the gas turbine exhaust without any supplementary heating. In the gas turbine, the compressor stage raises the air temperature from 293 K to 550 K before combustion which raises the temperature further to 1400 K. After expansion through the turbine, the gas temperature is reduced to 850 K. If the steam generating stage has an efficiency of 30%, and assuming the gas turbine may be described by an ideal Joule (Brayton) cycle, calculate the efficiencies of the gas turbine system and of the combined cycle.

4. What would be the Carnot limiting efficiency for a station operating at the same combustion temperature as in Question 3?

Chapter 7

NUCLEAR POWER GENERATION

7.1 Nuclear Energy: Equivalence of Mass and Energy

Nuclear power is derived from the conversion of mass into energy according to Einstein's equation relating the equivalence of mass and energy, $E = mc^2$. The energy bound up in mass is considerable due to the large value of the speed of light and hence, this represents a substantial source of power. This chapter sets out to describe the underlying physics and technology of nuclear power as it was developed during the mid-20th century. Starting with a brief discussion of the nucleus and the forces that hold it together, we will show how under certain circumstances, the nucleus can be split (i.e. a process called *fission*) by neutrons with a corresponding release of energy and how that process can be sustained and controlled in a nuclear reactor. We shall also review some of the technology of nuclear power systems before concluding with some consideration of the safety issues surrounding the disposal of nuclear waste and a brief account of the two main nuclear accidents to date — Three Mile Island and Chernobyl.

The nucleus of an atom is composed of neutrons and protons, collectively known as nucleons. The number of protons (Z) determines the element and the total number of nucleons (A) the atomic mass, the difference being the number of neutrons (N) of course. The notation used is (^A_Z Symbol) where 'Symbol' is the chemical symbol assigned to a given element in the periodic table. In this form the atomic number (Z) is redundant as a means of identification, but the atomic mass number (A) is not since most elements exist in more than one isotope, i.e. with different numbers of neutrons. For example, mercury (Hg, $Z = 80$) has 7 different stable isotopes ($A = 196, 198, 199, 200, 201, 202$ and 204). Ordinarily, mercury 200 (the most common isotope of Hg) would be denoted ^{200}Hg. Atomic mass is measured in *atomic mass units* (u) defined such that the mass of ^{12}C is exactly 12 u ($1u = 1.6605 \times 10^{-27}$ kg).

On the face of it, the nucleus should be unstable due to the Coulombic repulsion forces between protons, however, the fact that nuclei are evidently stable implies that there is a strong albeit short range force which dominates within the nucleus. This strong or hadronic force operates over ranges of the order of a few fm[a] and is related to the interactions between neutrons and protons. It is common to think of the nucleons in the nucleus as being confined within a deep potential well (Fig. 7.1).

---

[a]1 fm $= 10^{-15}$ m

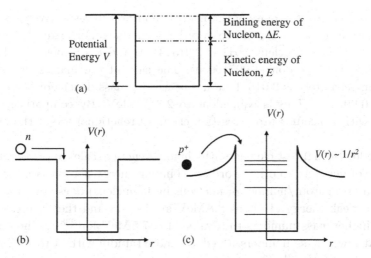

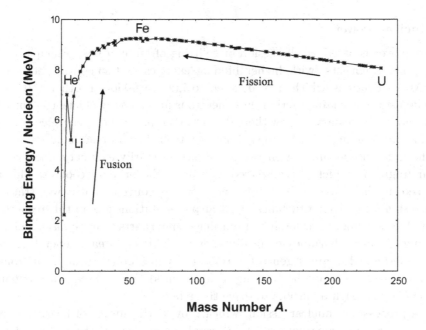

Fig. 7.1 Nuclear potential well (a) energies; (b) potential barrier for neutron which carries no charge; (c) Coulombic potential barrier for (charged) proton.

Fig. 7.2 Binding energy per nucleon.

With the exception of hydrogen, the atomic mass number is always less than the sum of the constituent nucleons. The mass deficit (Δm), as it is known, provides the energy for the nuclear potential well and the binding energy (ΔE) for the strong force:

$$A < (Nm_n + Zm_p) \quad \text{and} \quad \Delta E = \Delta mc^2 = (Nm_n + Zm_p - A) \times c^2 \qquad (7.1)$$

where m_n and m_p are the mass of the neutron and proton respectively. We can see that the mass deficits correspond to very large energies. For example, the atomic mass of a deuterium[1] nucleus (^{2}H) is 2.013231 u, and is composed of 1 neutron (1.008551 u) and 1 proton (1.007160 u). The sum of the masses of the proton and neutron separately is 2.015711 u and hence the mass deficit for the deuterium nucleus is 0.00248 u. This is equivalent to 2.3168 MeV. By comparison, energies associated with molecular processes (i.e. chemical reactions) are of the order of a few eV.

Except for the lightest elements, the binding energy scales approximately with the number of nucleons, in other words the binding energy per nucleon (i.e. binding energy for a given atom/number of nucleons in its nucleus) does not vary strongly. It reaches a peak energy of about 8.8 MeV at $A \sim 58$ and then decreases slowly with A at higher mass numbers to a value of ~ 7.5 MeV for ^{238}U. These are large energies. At lower mass numbers it reduces more rapidly with A to ^{12}C, and then varies strongly at lower A. The variation of binding energy per nucleon is shown in Fig. 7.2.

7.2 Nuclear Power

Nuclear power is based on the fission or splitting of large atoms, principally ^{235}U, into smaller fragments, with higher binding energies and some energetic or *fast* (> 10 MeV) neutrons which may be used to initiate fission in other ^{235}U atoms in a self-sustaining chain reaction. Fast neutrons are not very efficient at initiating fission and it is necessary to slow them down to *thermal* (< 1 eV) energies, a process known as *moderation*, in order to sustain a controlled chain reaction.

The difference in binding energy is released primarily as kinetic energy of the nuclear fragments which is transferred as heat to the primary coolant. This is in turn used as the source of heat in a more or less conventional power station to produce steam for driving turbines. Nuclear power stations are, in this respect, not greatly different from their fossil fuel burning counterparts that we discussed in the preceding chapter. Thermodynamically, they resemble the steam generating cycles where superheated steam is generated isobarically in a boiler (using heat from the reactor rather than from the burning of fossil fuels), and subsequently expanded isentropically through a turbine to drive the generator.

The processes of nuclear fission, which provide the source of heat, take place in the reactor core which houses the principal components of the fuel, moderator, control rods and coolant systems. The fuel is generally in the form of uranium metal or uranium dioxide ceramic pellets encased in cylindrical cans — fuel rods — embedded in the moderator, typically, water or graphite. Heat is removed from the core by a suitable coolant, but because of potential radiation leakage, this is not normally used as the working fluid to drive the turbines directly. Instead, the coolant is used to heat water and produce steam in a heat exchanger (boiler) which is used to drive the turbines (Fig. 7.3). However, there are serious constraints. Firstly, the

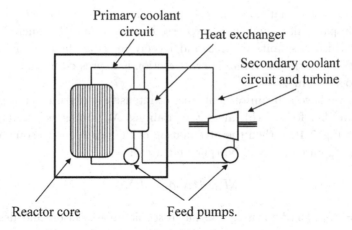

Fig. 7.3   Schematic diagram of a general reactor power system.

core temperature is limited, and secondly, the use of a secondary fluid means that the steam conditions are not ideal, resulting in complex and costly boiler/turbine systems with regenerative feed heating. We shall return to discuss these issues later in the chapter.

7.3 Neutron Dynamics

7.3.1 *Neutron cross sections*

There are three neutron reactions of consequence in the context of nuclear power:

(i)   *Radiative capture*: The target nucleus (X) absorbs the neutron, increasing the atomic mass number by 1 unit, is excited and subsequently decays by the emission of a gamma photon, i.e. $n + {}^{A}X \rightarrow {}^{A+1}X + \gamma$. Radiative capture of this type is often denoted (n,γ), and is important in nuclear reactors because it removes neutrons from the process, making it more difficult to sustain a chain reaction, and simultaneously producing harmful radiation.

(ii) *Scattering*: The target nucleus scatters the incident neutron, changing its velocity in the process. Scattering may be elastic, where kinetic energy is conserved, or inelastic, where some fraction of the neutron kinetic energy is absorbed by the target nucleus which becomes excited as a result. Scattering is sometimes denoted (n,n). Elastic scattering is the principal mechanism of moderation as fast neutrons transfer kinetic energy to the atoms of the moderator.

(iii) *Fission*: The target nucleus, usually ${}^{235}U$, absorbs the neutron and breaks up (or fissions) into two daughter nuclei and a number of neutrons. The daughter nuclei will be radioactive, decaying in turn to other radioactive species, some of which will have very long half lives.

The probability that any of the above reactions occur when a neutron encounters a particular nuclide is expressed in terms of the microscopic cross section, σ_i, which has units of area and loosely may be thought of as a 'target area'. Traditionally, the microscopic cross section is expressed in units of barns (b) equivalent to 10^{-28} m^2.

Suppose we have a beam of neutrons, of intensity I per unit area per second, incident on a thin foil of thickness t containing N_v atoms per unit volume as illustrated in Fig. 7.4(a), then the attenuation of the beam due to some interaction i (scattering, capture etc.) will be (per unit area):

$$\Delta I = ItN_v\sigma_i = It\Sigma_i \tag{7.2}$$

where Σ_i ($= N_v\sigma_i$) is known as the macroscopic cross section and describes the attenuation of a neutron beam as it passes through the foil. We can see this by considering the passage of a neutron beam of unit area through a small element dx at x in a 1-dimensional slab (Fig. 7.4b) where the neutrons are being scattered and/or absorbed out of the beam. If there are N_v nuclei per unit volume in the slab, then the attenuation will be

$$dI = -(\sigma_i N_v)\,I dx = -\Sigma_i I dx. \tag{7.3}$$

Integrating between the limits 0 and x

$$I(x) = I_0 \exp\left(-\Sigma_i x\right) \equiv I_0 \exp\left(-\frac{x}{\lambda_i}\right) \tag{7.4}$$

where I_0 is the beam intensity at $x = 0$.

Note that Σ has units of (1/length) and penetration through the slab by a distance $(1/\Sigma)$ reduces the beam intensity by a factor 'e'. The distance that a neutron may travel with a probability e that it will not encounter a nucleus, i.e. the mean free path, λ_i is therefore just the reciprocal of the macroscopic cross section $(1/\Sigma_i)$.

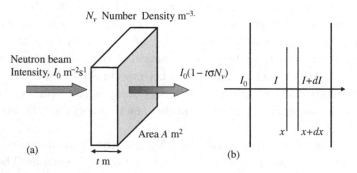

Fig. 7.4   (a) Attenuation of a neutron beam incident on a foil; (b) attenuation of the neutron by a differential element of the foil.

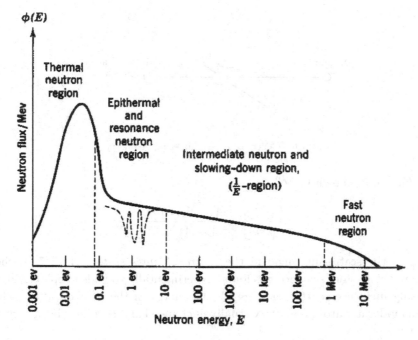

Fig. 7.5   Distribution of neutron energies in a typical thermal reactor.

The distribution of neutron energies within a reactor varies from the very energetic ($> 10\,\mathrm{MeV}$) neutrons produced at fission to fully thermalised neutrons ($< 1\,\mathrm{eV}$) in equilibrium with the surroundings. The neutrons are classified by their energies into fast, intermediate, epithermal and thermal (Fig. 7.5), to reflect the corresponding principal reactions with uranium. Thus, in the intermediate range (n,n) reactions predominate, while in the epithermal region (n,γ) reactions are dominant and neutrons are removed through absorption.

Only very energetic or *fast* neutrons can initiate fission in ^{238}U, the most common isotope, but such energetic neutrons quickly lose their energy and rapidly fall below the fission threshold for ^{238}U, making it difficult to sustain a chain reaction in ^{238}U alone. Although fast fission makes a small contribution to the reaction, it is fission in the less common isotope ^{235}U that is of greatest importance for nuclear power production. It is only the thermal neutrons that have a significant cross section for fission (σ_f) in ^{235}U, hence the need for moderation. This is achieved by the incorporation of additional materials within the reactor core that have a high cross section for scattering (σ_s) but a low capture cross section (σ_c) to minimise neutron losses.

7.3.2 *Scattering of neutrons*

In elastic scattering between a moderator atom of atomic mass number A and a neutron (mass $\approx 1\mathrm{u}$), the ratio of the final kinetic energy, E_1 to the incident kinetic

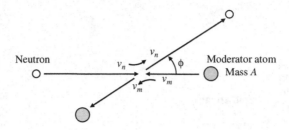

Fig. 7.6   Centre of mass scattering geometry.

energy E_0 can be shown to be[b]

$$\frac{E_1}{E_0} = \frac{1 + A^2 + 2A \cos \phi}{(1 + A)^2}$$

(7.5)

where ϕ is the scattering angle in the *centre of mass system* (Fig. 7.6), chosen because in this system scattering is found experimentally to be isotropic. Evidently, for grazing incidence scattering ($\phi \sim 0$), the energy of the neutron after collision with the moderator atom remains virtually unchanged and there has been very little moderation.

$$\frac{E_1}{E_0} = \frac{1 + A^2 + 2A}{(1 + A)^2}; \quad \text{i.e. } E_1 = E_0.$$

(7.6)

For a head-on collision ($\phi = 180°$), energy loss is a maximum

$$\left(\frac{E_1}{E_0}\right)_{\text{max}} = \frac{1 + A^2 - 2A}{(1 + A)^2} = \frac{(A - 1)^2}{(1 + A)^2} = \alpha$$

(7.7)

where α is the maximum possible energy loss. We can see by inspection of (7.7) that for moderators of large atomic mass number, α tends to unity (i.e. $E_1 \sim E_0$) and therefore low mass number moderators are preferred. In fact, for hydrogen ($A = 1$), all of the neutron energy would be transferred in a single head-on collision.

We can write for the maximum fractional energy loss following a single collision with a moderator atom

$$\left(\frac{\Delta E}{E_0}\right)_{\text{max}} = \left(1 - \frac{E_1}{E_0}\right)_{\text{max}} = (1 - \alpha).$$

(7.8)

Hence for a given moderator of mass number A the energy lost by the neutron per scattering event varies with the scattering angle in the range

$$0 < \left[\frac{\Delta E}{E_0}\right] < (1 - \alpha) \quad \text{for } 0 < \phi < \pi.$$

(7.9)

---

[b]This is a standard result in classical mechanics.

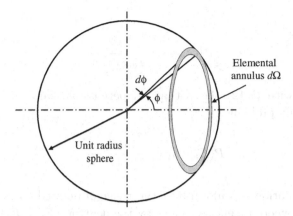

Fig. 7.7 Sphere of unit radius. Probability that a neutron will be scattered into an angle between ϕ and $d\phi$ is given by the ratio of the area $d\Omega$ of the elemental annulus to the total surface area 4π.

From the definition of α from (7.7) we get:

$$1 + \alpha = \frac{2\left(1 + A^2\right)}{\left(1 + A\right)^2} \quad \text{and} \quad 1 - \alpha = \frac{4A}{\left(1 + A\right)^2} \tag{7.10}$$

substituting into (7.5) we get the useful relationship between scattering angle and maximum energy loss:

$$\frac{E_1}{E_0} = \frac{1 + \alpha}{2} + \frac{1 - \alpha}{2} \cos \phi. \tag{7.11}$$

The probability $P(E)$ that a neutron will have a given energy E_1 after scattering is according to (7.5) and (7.11) proportional to $\cos(\phi)$ and consequently to the scattering angle. To determine the probability $P_\phi(\phi)$ that a neutron will be scattered through an angle between ϕ and $\phi + d\phi$, we define the unit radius sphere centred on the scattering point, O (Fig. 7.7).

As a fraction of the total surface area of the sphere, the area of the elemental annulus, $d\Omega$ is

$$P_\phi\left(\phi\right) d\phi = \frac{d\Omega}{4\pi} = \frac{2\pi \sin \phi d\phi}{4\pi} = \frac{1}{2} \sin \phi d\phi = -\frac{1}{2} d\left(\cos \phi\right). \tag{7.12}$$

Differentiating (7.11) with respect to ϕ[c]

$$dE_1 = E_0 \frac{\left(1 - \alpha\right)}{2} \sin \phi d\phi = E_0 \frac{\left(1 - \alpha\right)}{2} d\left(\cos \phi\right). \tag{7.13}$$

---

[c]Note dE_1 is a negative quantity as it represents an energy loss.

Hence

$$\frac{1}{2}d\left(\cos\phi\right) = \frac{dE_1}{E_0\left(1-\alpha\right)} \qquad (7.14)$$

and comparison with (7.12) shows that all permissible values of energy after a collision are equally probable:

$$P(E_1)dE_1 = \frac{dE_1}{E_0\left(1-\alpha\right)}. \qquad (7.15)$$

The energy distribution is uniform over the range of allowed energies, Fig. 7.8 i.e. between the case where no energy is lost by the neutron ($E_1 = E_0$) and maximum energy reduction ($E_1 = \alpha E_0$).

In the moderation process, we need to be able to predict the average number of collisions required for the majority of neutrons to thermalise. Neutrons will lose energy in discrete steps, but with the exception of the lightest elements (i.e. H), the fractional energy loss per collision is sufficiently small to model the problem as a continuous one, referred to as the continuous slowing down model (Fig. 7.9). The energy range (> 7 decades) can only be represented by a logarithmic scale and hence it is appropriate to consider the logarithmic energy decrement $-\Delta\log(E)$:

$$-\frac{dE}{E} = -d(\log E)\text{: i.e. } -\int_{E_o}^{E_1} d(\log E) = \log(E_0) - \log(E_1) \equiv \Delta\log(E). \quad (7.16)$$

For a sufficiently large number of collisions, we may calculate the average logarithmic decrement per collision, ξ as

$$\xi = \overline{\Delta\log(E)} = \overline{\log\left(\frac{E_0}{E_1}\right)} = \frac{\int_{\alpha E_0}^{E_0} \log\left(\frac{E_0}{E_1}\right) P(E)dE_1}{\int_{\alpha E_0}^{E_0} P(E_1)dE_1}. \qquad (7.17)$$

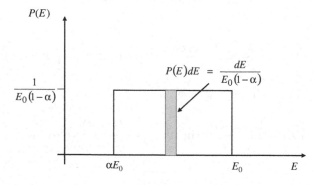

Fig. 7.8   Probability distribution of allowed energies after scattering.

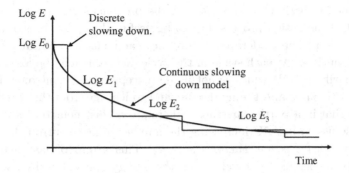

Fig. 7.9 Continuous slowing down approximation.

Substituting for $P(E_1)$ from (7.15) and changing the variable in (7.17) to $x = (E_1/E_0)$

$$\xi = -\frac{1}{(1-\alpha)} \int_\alpha^1 \log{(x)}\, dx = \frac{-1}{(1-\alpha)} [x \log{(x)} - x]_\alpha^1 = 1 + \frac{\alpha \log \alpha}{(1-\alpha)}. \qquad (7.18)$$

In terms of atomic mass number A, from (7.7)

$$\xi = 1 + \frac{(A-1)^2}{2A} \log \left(\frac{A-1}{A+1} \right). \qquad (7.19)$$

Clearly, for large A, $\xi \to 0$ as the argument of the logarithm tends to unity. In fact, for $A > \sim 10$, the average logarithmic decrement tends to a value of $(2/(A+(2/3)))$.

The average logarithmic decrement only gives the mean energy reduction for each scattering event, it provides no information as to the likelihood of a scattering event occurring. Thus, while hydrogen has a large limiting value of ξ the chances of a neutron encountering a hydrogen atom in a gas are relatively low. Alternatively, a high density solid with a modest ξ, might prove a better moderator material. We combine both requirements therefore in the *slowing down power* (SDP) defined as the product of ξ and the macroscopic cross section for scattering and use it as a figure of merit for the moderating material.

In fact, a more useful figure of merit is the moderator ratio (MR) defined as the ratio of the SDP to the macroscopic cross section for absorption, since, as it will become evident, we need to avoid the loss of neutrons to non-fissioning processes.

$$\text{SDP} = \xi \Sigma_S; \quad \text{MR} = (\xi \Sigma_s)/\Sigma_a. \qquad (7.20)$$

7.4 Reactor Physics

7.4.1 *The nuclear chain reaction*

The essence of producing nuclear power in a controlled manner is to sustain a chain reaction such that the supply of thermal neutrons is just sufficient to replenish those consumed in fission. As discussed in the previous section, fast neutrons are a by-product of the fission process, but to be useful, they must be slowed down to thermal energies, i.e., be moderated, before they can be used to initiate the fission of another ^{235}U nucleus. At each stage in the cycle, neutrons may be lost to radiative capture primarily in ^{238}U which is not fissile (except to fast neutrons), but also in surrounding structures and the moderator and to leakage from the reactor core.

Suppose that in a particular reactor there are n_0 fast neutrons, and a fraction of these cause fission in ^{238}U increasing the number of fast neutrons by a factor ε, known as the *Fast Fission Factor* $(\varepsilon > 1)$. This group of fast neutrons now numbering (εn_0) begins to travel across the energy spectrum (Fig. 7.5) as it is slowed down by collisions with the moderator. At epithermal energies, the neutrons encounter strong (n,γ) absorption resonances in ^{238}U and a significant proportion are lost as a result. If the *Resonance Escape Probability* (i.e. of not being absorbed) is p_e (< 1), then the total number of neutrons reaching thermal energies will be $(p_e \varepsilon n_0)$. Of these, only a fraction f, the *Thermal Utilisation Factor*, given by the ratio of the macroscopic absorption cross section for fuel (Σ_{aU}) to the total macroscopic absorption cross section (Σ_{aTotal}), will be absorbed by ^{235}U to cause fission

$$f = \frac{\Sigma_{aU}}{\Sigma_{aTotal}}. \tag{7.21}$$

If the average number of fast neutrons produced per fission is η then the total number of new fast neutrons ready to restart the cycle is $(\eta f p_e \varepsilon n_0)$. We define the *Infinite Reproduction Constant* (k_∞) as the ratio of the number of neutrons left at the end of the cycle to the number at the beginning, i.e.

$$k_\infty = \frac{\eta f p_e \varepsilon n_0}{n_0} = \eta f p_e \varepsilon. \tag{7.22}$$

The ^{238}U absorption cross section in the epithermal part of the spectrum varies strongly and in a very complex way, but it is can be shown that

$$p_e = \exp\left\{-\frac{N_0}{\xi \Sigma_s} \int_{E_1}^{E_0} \frac{\sigma_a}{1 + N_U \sigma_a / \Sigma_s} \frac{dE}{E}\right\} = \exp\left\{-\frac{N_0}{\xi \Sigma_s} \int_{E_1}^{E_0} (\sigma_a)_{eff} \frac{dE}{E}\right\} \tag{7.23}$$

where N_U is the number density of uranium nuclei.

The integral in (7.23) is known as the *Effective Resonance Integral* and the integrand $(\sigma_a)_{eff}$ the *Effective Absorption Cross Section*. The value of the effective resonance integral and effective absorption cross sections will depend on the specific design of a given reactor system and must be determined empirically for each case. Equation (7.23) gives the probability that a neutron will travel from energy E_0 to E_1 without being absorbed. It provides no information on the detailed motion of the neutron, which will be random.

Equation (7.22), sometimes referred to as the *Four Factor Formula,* describes neutron reproduction in the absence of leakage from the reactor core. It corresponds to a reactor of infinite extent and this is why the suffix '∞' is added to the reproduction constant k_∞. In a reactor of finite dimensions, neutron leakage from the core represents a significant loss to the system and this must be accounted for.

Conventionally, we separate the leakage into fast and thermal neutron components due to differences in the physical processes for the two classes and define fast and thermal non-leakage factors, (l_f) and (l_{th}) respectively. The infinite reproduction (k_∞) must be multiplied by these factors to obtain the actual or *Effective Reproduction Constant,* (k_{eff})

$$k_{eff} = k_\infty l_f l_{th} = (\eta f p_e \varepsilon)\, l_f l_{th}. \tag{7.24}$$

The magnitude of k_{eff} determines the self multiplication rate for the neutrons in the reactor assembly.

(1) $k_{eff} < 1$: more neutrons are being consumed or lost than are being produced, the reactor is said to be sub-critical and the system will shut itself down.

(2) $k_{eff} = 1$: the neutron population is stable with neutron consumption being exactly balanced by production and the reactor is said to be critical.

(3) $k_{eff} > 1$: the reactor is super-critical and the neutron population is increasing.

A real reactor must be able to operate safely in all three regimes: it must be super-critical during start-up, critical in continuous operation; and sub-critical during close down. This is done by inserting 'control' rods of a material with a high neutron absorption cross section (typically Cd) to, in effect, adjust $(\sigma_a)_{eff}$.

We treat thermal and fast neutron leakage differently because the physics is different. In the case of thermal neutrons, the neutrons are in equilibrium with their surroundings, i.e. at the same temperature as the reactor core and may be considered to be mono-energetic. On the other hand, fast neutrons are continuously changing their energy as they slow down and are therefore not mono-energetic. We define the *Slowing Down Density* $q(E)$ as the number of neutrons per unit volume that transit past energy E in unit time. The slowing down density at thermalisation, q_{th}, is the number density of neutrons 'arriving' at thermal energies in unit time. The leakage of both groups of neutrons is determined by the respective diffusion processes. We shall first consider the diffusion of thermal neutrons.

7.4.2 *Thermal neutron diffusion and leakage*

The thermal neutron density at some point (x, y, z) in the reactor will depend on the balance between production (i.e. the slowing down density at thermal energies (q_{th})), absorption and leakage. We express this in differential equation form as the time rate of change in the neutron population

$$\frac{\partial n}{\partial t} = \text{production } (q_{th}) - \text{absorption} - \text{leakage}. \tag{7.25}$$

The rate of absorption is given by the product of the macroscopic absorption cross section (Σ_a) and the number of neutrons passing through unit area in unit time (nv), v being the mean neutron velocity.

Leakage of neutrons is due to diffusion out of the reactor and is therefore governed by Fick's First law of Diffusion.

$$\underline{J}_{dif} = -D\nabla n \tag{7.26}$$

where D is the diffusion constant and is given by kinetic theory as $D = 1/3(v\lambda_{tr})$, where λ_{tr} is the transport mean free path.[d] Conservation of mass requires that the rate of change of the neutron density due to neutron leakage must equal the divergence of the neutron current density

$$\nabla \cdot \underline{J}_{dif} = -\frac{\lambda_{tr}v}{3}\nabla \cdot \nabla n = -\frac{\lambda_{tr}v}{3}\nabla^2 n. \tag{7.27}$$

In steady state, $(\partial n/\partial t) = 0$ and thermal neutron production is exactly equal to the losses due to absorption and leakage. Substituting (7.27) in (7.25), and rearranging the resulting equation

$$\nabla^2 n - \frac{3}{\lambda_a \lambda_{tr}}n + \frac{3q_{th}}{v\lambda_{tr}} = 0 \tag{7.28}$$

where we have replaced the macroscopic absorption cross section by the mean free path for absorption ($\Sigma_a = 1/\lambda_a$; (Sec. 7.3.1)). Equation (7.28) is the *Steady State Equation for Thermal Neutron Diffusion* and describes how the thermal neutron population would be distributed throughout a reactor core of given geometry under steady state conditions.

Solutions to (7.28) depend on the geometry. The simplest case (somewhat unrealistic) is for a point source of neutrons at the origin in a homogeneous isotropic medium. Since $q_{th} = 0$ everywhere except at the source, (7.28) reduces to

$$\nabla^2 n - \frac{3}{\lambda_a \lambda_{tr}}n \equiv \nabla^2 n - \frac{1}{L_{th}^2}n = 0; \quad \text{where } L_{th}^2 = \frac{\lambda_a \lambda_{tr}}{3}. \tag{7.29}$$

L_{th} is the *thermal neutron diffusion length* and is the average 'line-of-sight' distance a neutron travels between its origin as a thermal neutron and its eventual absorption. In solving an equation like (7.28), it is sensible to express the Laplacian in terms of a co-ordinate regime that reflects the inherent symmetries of the problem in question. In the present example, where we have a point source at the origin of an otherwise

---

[d]Neutrons tend to scatter more strongly in the forward direction and thus, a neutron will on the average move away from its point of origin. The transport mean free path is the effective increased mean free path due to preferred forward scattering.

uniform isotropic medium, the prudent choice is to use a spherical polar co-ordinate system

$$\nabla^2 n - \frac{n}{L_{th}^2} = \frac{1}{r^2}\frac{\partial}{\partial r}\left(r^2\frac{\partial n}{\partial r}\right) - \frac{n}{L_{th}^2} = 0. \tag{7.30}$$

Changing the variable to $R = nr$; or $n = R/r$ and substituting in (7.30):

$$\frac{1}{r^2}\frac{d}{dr}\left(r^2\frac{1}{r}\frac{dR}{dr} - R\right) - \frac{R}{rL_{th}^2} = \frac{1}{r^2}\left(r\frac{d^2R}{dr^2} + \frac{dR}{dr} - \frac{dR}{dr}\right) - \frac{R}{rL_{th}^2} = \frac{d^2R}{dr^2} - \frac{R}{L_{th}^2} = 0 \tag{7.31}$$

which is a simple second order ordinary differential equation having the solution:

$$R = A\exp\left[\frac{r}{L_{th}}\right] + B\exp\left[-\frac{r}{L_{th}}\right] \tag{7.32}$$

where A and B are constants to be determined by the boundary conditions, namely that $n \to 0$ as $r \to \infty$ and that total absorption = total thermal neutron production (q_{th}). The first condition indicates that $A = 0$, otherwise the first exponential term would increase without limit. The second condition requires

$$q_{th} = \int_0^\infty v n\Sigma_a dV = \int_0^\infty v n\Sigma_a 4\pi r^2 dr \tag{7.33}$$

where we have taken dV to be the volume of a shell centred on the origin. Substituting for n by B/r and integrating (using (7.29)) gives the final solution:

$$n(r) = \frac{q_{th}\Sigma_a}{4\pi v L_{th}^2}\frac{1}{r}\exp\left(-\frac{r}{L_{th}}\right) = \frac{3q_{th}}{4\pi v\lambda_{tr}}\frac{1}{r}\exp\left(-\frac{r}{L_{th}}\right). \tag{7.34}$$

As one would expect, the neutron density varies inversely with distance from the source.

7.4.3 *Fast neutron diffusion and leakage*

The fast neutrons are not in equilibrium with the reactor, instead, they display a spectrum of energies as they slow down. Consequently, we have to consider not only their diffusion in real space but also their passage through energy space. In other words, a proper description of the fast neutrons must include both their energy and spatial distributions.

We may represent the neutron flow in a space — energy co-ordinate diagram such as Fig. 7.10, in which a suitable spatial dimension is plotted on the vertical axis against energy on the horizontal axis. We consider an element on the space-energy diagram between energies, E and $E + \Delta E$ and make the assumption[e] that the only

---

[e]Reactors are generally designed so that this is the case, i.e. fuel-moderator geometries are designed to ensure that during the slowing down process neutrons are confined to the moderator (i.e. $p_e \sim 1$).

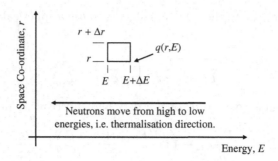

Fig. 7.10   Space co-ordinate — energy diagram for fast neutrons during slow down.

physical processes of consequence are diffusion and scattering (there are no sources or sinks of neutrons). Neutrons entering from the right do so at higher energies, and are thus moving more rapidly than those leaving from the left — this is why fast neutrons must be treated different from thermal ones. To ignore this would violate continuity of matter — neutron influx must equal neutron outflow (7.27).

$$q\left(E + \Delta E\right) - q\left(E\right) = \frac{\partial q}{\partial E}\Delta E = -\frac{\lambda_{tr}v}{3}\nabla^2 n. \tag{7.35}$$

The slowing down density $q(E)$ is the number of neutron per unit volume that pass energy E in unit time. This must be equal to the scattering rate $(v\Sigma_s)$ and the neutron density in the energy interval ΔE. Using (7.16) and (7.17)

$$q(E) = n(E)\Delta E v \Sigma_s = n(E)E\xi v \Sigma_s. \tag{7.36}$$

Differentiating twice and rearranging

$$\nabla^2 n = \frac{\nabla^2 q}{v\Sigma_s \xi E}. \tag{7.37}$$

Substituting for $\nabla^2 n$ in (7.35) gives us an equation that relates the energy and spatial dependence of the slowing down density and thus provides us with a full description of the neutron and energy distribution within the reactor.

$$\nabla^2 q = \frac{\partial q}{(-\lambda_{tr}\lambda_s/3\xi)(dE/E)} \equiv \frac{\partial q}{\partial \tau} \quad \text{where } d\tau = -\frac{\lambda_{tr}\lambda_s}{3}\frac{1}{\xi}\frac{dE}{E}. \tag{7.38}$$

The variable 'τ' is known as the *Fermi Age* and (7.38) is more commonly written as

$$\nabla^2 q - \frac{\partial q}{\partial \tau} = 0 \tag{7.39}$$

in which form it is know as the *Fermi Age Equation*. It should be noted that τ is not a time, its units are actually (length)^2. The Fermi Age at neutron birth is

$\tau(E = E_0) = 0$ and integrating from neutron birth is $(E = E_0)$ to energy E^{f}

$$\int_E^{E_0} d\tau = 0 - \tau(E) \equiv -\frac{\lambda_{tr}\lambda_s}{3} \int_E^{E_0} \frac{1}{\xi} d(\log E) = -\frac{\lambda_{tr}\lambda_s}{3} \frac{\log(E_0/E)}{\xi}. \quad (7.40)$$

The term $(1/\xi)(\log(E_0/E))$ is the average number of times a neutron is scattered as its energy is reduced from $E_0 \to E$ and if the average number of collisions required to slow a neutron to thermal energy is λ_f, then by analogy with (7.29) we can define the *Fast Diffusion Length* to be

$$\tau(E_0 \to E_{th}) = \frac{\lambda_{tr}(\lambda_f \lambda_s)}{3} \equiv L_f^2. \quad (7.41)$$

The bracketed term $(\lambda_f \lambda_{tr})$ corresponds to the average total (i.e. zigzag) path length traced out by a neutron during its slowing down phase, Fig. 7.11. It is directly analogous to λ_a in (7.29), in that it defines the life-time of the neutron in the particular regime.

Solutions to the Fermi Age equation also depend on the reactor geometry, but for a point source emitting neutrons isotropically into a uniform medium at constant rate (q_0), the slowing down density at radial distance r is the Gaussian

$$q(r) = q_0 \frac{\exp(-r^2/4\tau)}{(4\pi\tau)^{\frac{3}{2}}}. \quad (7.42)$$

Equation (7.42) gives the neutron distribution at a given Fermi age. It indicates that the neutron density is always greatest at the origin, but as the Fermi Age increases (energy decreases) the distribution broadens out and the peak value diminishes, until $\tau = \tau_0$, the Fermi Age at thermalisation.

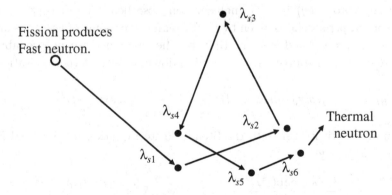

Fig. 7.11   Trajectory of a neutron during moderation from fast to thermal energies.

---

^{f}Note that $\frac{dE}{E} \equiv d(\log E)$.

7.4.4 *Critical equation*

The Fermi Age Equation (7.39) is in four dimensions; the three normal spatial dimensions and the Fermi Age as a fourth. This makes the solution more complicated, but the equation may be readily solved if the variables are separable. If they are, the slowing down density may be expressed as the product of two functions, one in the space variable (r) only and the other in the Fermi Age (τ) only, such that

$$q(r, \tau) = R_r(r) T_\tau(\tau). \tag{7.43}$$

It follows that

$$\nabla^2 q = T_\tau \nabla^2 R_r \quad \text{and} \quad \frac{\partial q}{\partial \tau} = \frac{\partial T_\tau}{\partial \tau} R_r. \tag{7.44}$$

Substitution into the Fermi Age Equation gives

$$T_\tau \nabla^2 R_r - \frac{\partial T_\tau}{\partial \tau} R_r = 0 \quad \text{or} \quad \frac{\nabla^2 R_r}{R_r} = \frac{1}{T_\tau} \frac{\partial T_\tau}{\partial \tau}. \tag{7.45}$$

If we consider the second equation, then it is clear that the left-hand side is a function of the spatial co-ordinate (r) only and the right-hand side a function only of the Fermi Age (τ). This is only possible if they are both equal to a constant, say $(-B^2)$.

$$\frac{1}{T_\tau} \frac{\partial T_\tau}{\partial \tau} = -B^2 \text{ which has the solution } T_\tau(\tau) = T_\tau(0) \exp(-B^2 \tau). \tag{7.46}$$

Substituting back into (7.43), we have, at a given location (r) in the reactor

$$q(r, \tau) = R_r(r) T_\tau(0) \exp(-B^2 \tau) = q_0 \exp(-B^2 \tau) \tag{7.47}$$

where q_0, the slowing down density at $\tau = 0$, is the production rate of fast neutrons equal to $(\varepsilon f \eta)$ per thermal neutron absorbed. The density of neutrons being absorbed per second is given by $(nv\Sigma_a)$, and since some of these are absorbed while slowing down, we should multiply by the resonance escape probability, p_e. The thermal neutron source, i.e. the slowing down density at thermalisation is

$$q_{th} = (v\Sigma_a)(p_e \varepsilon f \eta) n \exp(-B^2 \tau) = \left(\frac{v}{\lambda_a}\right) k_\infty n \exp\left(-B^2 \tau\right). \tag{7.48}$$

We can now substitute this into the Steady State Equation for Thermal Neutron Diffusion (7.28) to get

$$\nabla^2 n - \frac{1}{L_{th}^2} n + \frac{3}{\lambda_{tr}} \frac{v k_\infty \exp(-B^2 \tau)}{\lambda_a} n = 0 \quad \text{or} \quad \nabla^2 n + \frac{k_\infty \exp(-B^2 \tau) - 1}{L_{th}^2} n = 0. \tag{7.49}$$

The slowing down density, q_{th} at thermal energies must also satisfy the Fermi Age Equation and consequently, we substitute (7.48) into (7.39) with the conditions (7.45)

$$\nabla^2 q_{th} - \frac{\partial q_{th}}{\partial \tau} = (v\Sigma_a)k_\infty \exp(-B^2\tau)\nabla^2 n + (v\Sigma_a)k_\infty B^2 \exp(-B^2\tau)n = 0$$

$$(v\Sigma_a)k_\infty \exp(-B^2\tau)\{\nabla^2 n + B^2 n\} = 0 \quad \text{or} \quad \nabla^2 n + B^2 n = 0. \tag{7.50}$$

Equating (7.49) and (7.50), we get the important result:

$$\frac{k_\infty \exp(-B^2\tau)}{1 + L_{th}^2 B^2} = 1 \equiv k_{eff}. \tag{7.51}$$

This is the *Critical Equation* and the right-hand side is effectively k_{eff} as indicated. The reactor is just critical when $k_{eff} = 1$ as previously discussed, and for a given k_∞, this is crucially dependent on τ, L_{th} and B. In (7.41) and (7.29) we identified τ and L_{th} as the fast and thermal neutron diffusion lengths respectively, where B is known as the *Buckling*. The diffusion lengths are defined as the mean distances travelled away from the point of production over the lifetime of the neutron; for fast neutrons this is the time between production at fission and thermalisation, while for thermal neutrons it is the time between thermalisation and final absorption. The question is whether the diffusion lengths are significant in relation to the size of the reactor core. If the reactor core is small, then the probability of leakage is correspondingly increased and it will be more difficult to maintain the condition $k_{eff} = 1$. The effects of reactor size in relation to the diffusion lengths is included in the buckling, which always appears in (7.51) as a product with a diffusion length. In other words, it *scales* the particular diffusion length. To a first approximation, B is inversely proportional to the reactor size, and consequently for an 'infinite' reactor, $k_{eff} = k_\infty$.

The geometric aspects of the buckling in (7.50) can be made explicit by writing the equation as the ratio of the second spatial derivative of the neutron density to the neutron density. This defines the 'curvature' or 'buckling' of the neutron density, hence the name

$$\frac{\nabla^2 n}{n} = -B^2. \tag{7.52}$$

In this form, B is intimately related to the reactor geometry and is therefore sometimes referred to as the *Geometric Buckling*, B_G. Similarly, the buckling in (7.51) is often referred to as the *Material Buckling*, B_M, since it is primarily determined by the material or physical properties of the reactor. Implicit in the discussion has been the assumption that $B_G = B_M$, which is another way of defining the condition for critical operation.

- $B_G < B_M$ Reactor is supercritical: in effect there is not enough neutron leakage from the core, and so neutron levels will rise inexorably unless control measures are taken to reduce the neutron population, i.e. insertion of control rods.
- $B_G = B_M$ Reactor is just critical: the desired operational regime.

- $B_G < B_M$ Reactor is sub-critical: leakage is too great, and the neutron population will decay away, unless steps are taken to increase thermal neutron production.

Geometric buckling is a function of the reactor shape and (7.52) has to be solved for the particular reactor in question. In practice, reactors are complicated structures and (7.52) cannot be solved analytically. However, as an illustrative example, consider the case of a parallelpipe-shaped homogenous reactor core with sides of length $x = a$, $y = b$, $z = c$, and we shall take the origin at the centre of the core (i.e sides are at $x = a/2$; $y = b/2$; $c = z/2$). Assume that the neutron density at the origin is a maximum and is zero at the edges. Equation (7.52) becomes

$$\frac{\partial^2 n_x}{\partial x^2} + \frac{\partial^2 n_y}{\partial y^2} + \frac{\partial^2 n_z}{\partial z^2} + nB_G^2 = 0 \quad \text{and}$$

$$\text{solution } n\left(x, y, z\right) = n_0 \cos\left(\frac{\pi x}{a}\right) \cos\left(\frac{\pi y}{b}\right) \cos\left(\frac{\pi z}{c}\right)$$

where n_0 is the neutron density at the centre of the core. Substituting back for $n(x,y,z)$

$$B_G^2 = \left\{\frac{\pi}{a}\right\}^2 + \left\{\frac{\pi}{b}\right\}^2 + \left\{\frac{\pi}{c}\right\}^2. \tag{7.53}$$

7.4.5 *Reactor kinetics*

Thus far in the analysis of a nuclear reactor, we have only discussed the steady state operation of the reactor, and we have also assumed that the only neutron source is due to the fission of ^{235}U. We have ignored the fact that the fission products are in general radioactive and some will decay by emission of a neutron, effectively adding to the neutron economy. Neutrons produced by fission are referred to as *prompt neutrons*, since they are emitted at the instant of fission. Those produced by radioactive decay of fission products are emitted sometime later and for obvious reasons are known as *delayed neutrons*. It turns out that the delayed neutrons are crucial to the operation and control of a nuclear reactor.

In the non-steady state case, the right hand side of (7.28) $\neq 0$. In fact equation (7.28) was derived from (7.25) under the *assumption* of steady state. For the non-steady condition, we have to start again from (7.25), substituting term by term for the components of the equation i.e.

$$\frac{\partial n}{\partial t} = q_{th} - \frac{v}{\lambda_a}n + \frac{\lambda_{tr}v}{3}\nabla^2 n. \tag{7.54}$$

Replacing q_{th} using (7.48) and rearranging

$$\frac{3}{v\lambda_{tr}}\frac{\partial n}{\partial t} = \frac{3}{v\lambda_{tr}}\frac{\lambda_a}{\lambda_a}\frac{\partial n}{\partial t} = \frac{\bar{t}}{L_{th}^2}\frac{\partial n}{\partial t} = \frac{k_\infty \exp\left(-B^2\tau\right) - 1}{L_{th}^2}n + \nabla^2 n \tag{7.55}$$

where $\bar{t}$ is the mean life time of a thermal neutron (i.e. mean free path, λ_a/average speed, v). For small departures from criticality, the neutron density and distribution will be close to the critical case and we may replace $\nabla^2 n$ in (7.55) by $-B^2$ (7.50). Multiplying through by L_{th}^2

$$\frac{\bar{t}}{n}\frac{\partial n}{dt} = -L_{th}^2 B^2 + k_\infty \exp(-B^2\tau) - 1 \quad \text{and} \quad \div \left(1 + L_{th}^2 B^2\right) \text{ gives}$$

$$\frac{\bar{t}}{1 + L_{th}^2 B^2}\frac{1}{n}\frac{\partial n}{\partial t} = \frac{k_\infty \exp\left(-B^2\tau\right)}{1 + L_{th}^2 B^2} - 1 \equiv k_{eff} - 1. \tag{7.56}$$

The term $(k_{eff} - 1)$ is the *Excess Reactivity*, (Δk) and $(\bar{t}/(1 + L_{th}^2 B^2))$ is the mean thermal neutron lifetime with leakage, t_0. Equation (7.56) gives the time dependence for the neutron population

$$\frac{\partial n}{\partial t} = \frac{\Delta k}{t_0}n \rightarrow n\left(t\right) = n\left(0\right) \exp\left(\frac{\Delta k}{t_0}t\right) = n\left(0\right) \exp\left(\frac{t}{T_R}\right) \tag{7.57}$$

where T_R is the *Reactor Period*, and is the time required for the neutron population to change by a factor of e (it is sometimes referred to as the e-folding period).

Mean lifetimes for prompt thermal neutrons (t_0) depend on the particular reactor design but are typically of the order of a few ms. Thus, for a reactor with only prompt neutrons, a small excess reactivity of say 0.2% would give a reactor period of ~ 0.5 s and the neutron population would increase by a factor of over 22,000 in about 5 s which would be very difficult to control.

Six groups of delayed neutrons have been identified according to half life, varying from about a third of a second to over a minute and constitute 0.64% of the fission neutron population. They are generated when radioactive fission products decaying by β^- emission leave behind daughter nuclei in highly excited states which subsequently relax by the emission of a neutron. Determining the average lifetime for all the neutron groups including the prompt neutrons leads to a value of ~ 0.1 s. The delayed neutrons have increased the mean lifetime and hence, for a given Δk, the reactor period by about two orders of magnitude. In terms of our earlier example, this would mean that the neutron population increased by a much more controllable factor of 1.1 in 5 seconds.

7.5 Reactor Systems

7.5.1 *Materials constraints*

The fundamental requirement in the steady state operation of a reactor is to maintain criticality, to ensure that each neutron consumed or lost is replaced. In addition to leakage and resonance absorption in ^{238}U, there will be significant absorption losses in other component materials, not least in the fuel cladding, moderator and coolant materials. Any shortfall in the neutron balance must be addressed by increasing the ratio of ^{235}U to ^{238}U in the fuel, a process known as *enrichment* and which results in an increase in the thermal utilisation factor, f. Natural uranium is composed of 0.71% ^{235}U, and while increasing this proportion (enrichment) is possible, it is expensive. As a result, the design of early reactor systems was often constrained by the need to use natural uranium.

Ideally, the moderator should have a low atomic mass number (7.7), but to avoid undue neutron loss, it should also have a low absorption cross section. With unit atomic mass, hydrogen should be the ideal moderator, but unfortunately, the absorption cross section is high (0.3 barns) and it can only be used with enriched fuel. Deuterium, with both a low absorption cross section (4.6×10^{-3} barns) and a low mass number, is an ideal but expensive moderator that can be used with natural uranium fuelled reactor systems, as can high purity graphite ($\sigma_a = 5 \times 10^{-3}$ barns). The higher mass number of graphite means that many more scattering events are required to slow the neutrons down and reactors are designed so that during the moderation process, neutrons are mainly confined to the graphite until they have reached thermal energies.

The fuel must be clad to prevent reaction with the coolant and the release of radioactive fission products. The fuel elements may be uranium metal or ceramic pellets of uranium dioxide (UO_2), uranium oxycarbide and less commonly, thorium oxide. The pellets are typically coated with successive layers of pyrolytic carbon and silicon carbide. Uranium metal undergoes a violent chemical reaction when it comes into contact with hot water and melts at the comparatively low temperature of 1132°C, and can therefore only be used in low temperature systems, limiting the available thermodynamic efficiency (Sec. 6.2). The ceramic fuels can be operated at higher temperature and do not react chemically with water, however, they have lower thermal conductivities. Ceramic fuels are being increasingly used as the fuel in modern power reactor systems. Typical fuel can materials are Mg-alloy and Zr alloys which have low neutron absorption, and stainless steel (due to its strength and corrosion resistance), although it can only be used with enriched uranium.

7.5.2 *Magnox gas cooled reactor*

The world's first commercial nuclear power station was the UK Magnox station at Calder Hall.[1] Commissioned in 1956 with an anticipated life-span of about twenty-five years, it was finally taken out of service in March 2003 after operating for nearly twice its expected life-span. It utilized a graphite moderated, CO_2 cooled reactor with natural uranium fuel elements encased in Magnox, the specially developed magnesium-aluminium alloy after which the reactor design was named. In total, 25 Magnox reactors delivering 4.8 GWe were installed worldwide, of which 22 (3.8 GWe) were located in the UK, many of which are still operational. Calder Hall delivered 200 MWe of electrical power from a suite of four reactors operating at a rather low overall thermal efficiency of 19%. The last Magnox station to be built (Wylfa in Wales) was capable of providing 1.1 GWe from two reactors at a thermal efficiency of $\sim 31\%$.

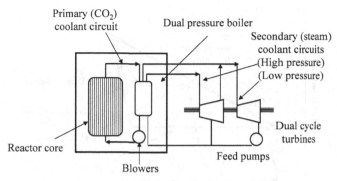

Fig. 7.12 Schematic diagram of Magnox reactor system, illustrating dual pressure cycle steam cycle.

The Magnox alloy used for cladding the uranium fuel was developed to allow the use of natural uranium and as an alternative to Al, which reacts with uranium to form a brittle compound. Magnox (an alloy of Mg with Al (1%) and Be (0.05%)) proved to be an acceptable compromise for low temperature operation ($< 600\ ^0$C). Heat is transferred to the boilers using CO_2 as the primary cycle coolant at pressures of ~ 280 MPa (in more advanced systems) at an outlet temperature of $\sim$410°C. This limits the operating range with the result that the turbine conditions in the secondary steam circuit are not ideal and to improve the steam cycle efficiency, Magnox systems used a dual pressure cycle system (Fig. 7.12). Some of the steam is produced at high pressure and supplied to a high pressure turbine, the rest generated at a lower pressure, mixed with the reduced pressure steam from the high pressure turbine outlet and supplied to a second low pressure turbine.

The use of a dual cycle pressure system improves the steam cycle efficiency by reducing the temperature difference between the CO_2 and the steam, i.e. increasing the average temperature at which heat is transferred (Fig. 7.13).

7.5.3 *Advanced Gas Cooled Reactor*

The advanced gas-cooled reactor (AGR) was intended to be a much more efficient second generation gas-cooled reactor. The main restriction on the efficiencies of the early Magnox systems was the low reactor temperature, dictated by the use of uranium metal and Magnox. These were replaced by ceramic UO_2 (enriched by 3%) and stainless steel respectively. The fuel cans were inserted in graphite sleeves, which were then assembled to form the core. Carbon dioxide was again used as the primary coolant at pressure and temperature of ~ 400 MPa and ~ 650°C respectively, considerably increased compared to the Magnox reactors.

Among the unique features of the AGR design are the housing of the reactor, heat exchangers and circulators within a single pre-stressed concrete enclosure, (the reactor core is contained in an inner pressure vessel) and the use of a mixed

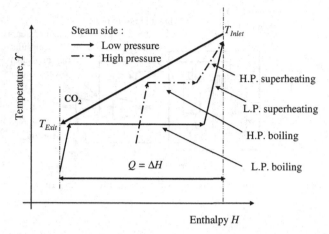

Fig. 7.13   Temperature-enthalpy diagram for dual pressure steam cycle used in Magnox reactor systems.

coolant flow regime. Some of the CO_2 is directed downwards through the core (re-entrant flow) to maintain the graphite and surrounding structures close to the inlet temperature. This is then mixed with the rest of the boiler exhaust to flow upwards through the fuel channels (Fig. 7.14).

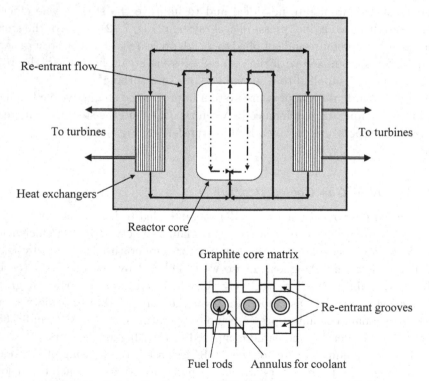

Fig. 7.14   Schematic diagram of AGR illustrating re-entrant flow.

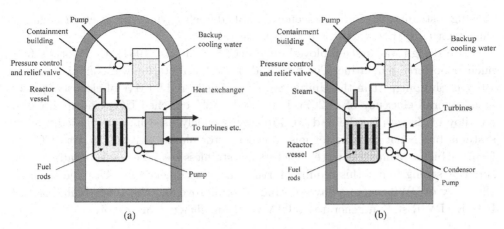

Fig. 7.15   Schematic diagrams of (a) PWR reactor and (b) BWR reactor.

The first AGR system was commissioned in 1963, and since then a total of 14 reactor units have been installed, each designed to generate 660 MW at a thermal efficiency of $\sim 40\%$. The high temperature and pressures permit the use of a single pressure turbine cycle i.e. operation under similar conditions to a standard steam generating fossil fuel station.

7.5.4 *Pressurised Water Reactor*

With several hundred units installed throughout the world, the pressurised water reactor (PWR) is the most widely used of the various power reactor systems.[2] It is illustrated schematically in Fig. 7.15(a). An American design, the PWR uses light water (H_2O) as both coolant and moderator. To compensate for neutron absorption in light water, it uses enriched ($\gtrsim 1\%$) uranium fuel, although due to the high moderating power of hydrogen, the reactor need not be as large as its graphite counterpart. To prevent the coolant/moderating water to become boiling while raising its temperature, the reactor and primary circuit has to be pressurised. However, the water must be maintained below the critical point ($T = 647.3$ K, $p = 22.1$ MPa), and thus it places an upper limit on the operating conditions. Typically, a PWR station would operate at pressures of 5 to 15 MPa and temperatures of $\sim 300°$C, with a single cycle secondary heat transfer system.

The primary and secondary coolant circuits employed in the PWR design effectively confines hazardous material to within the containment building under normal operation. However, the design suffers from one serious flaw. In the event of a sudden loss of coolant, the reactor could overheat catastrophically. Although the reactor would automatically shut down due to the loss of moderator, the delayed neutrons — so important in controlling normal operation (see Sec. 7.4.5) — would now act to slow the close down process, resulting in excessive heat production in the interim. Internal structures could melt and collapse into any remaining water causing a steam explosion — a scenario dubbed 'meltdown'. As a safeguard against this eventuality, the reactor is immersed in water.

The design was first employed on the USS Nautilus (1954), the World's first nuclear powered submarine. The first land-based PWR station was commissioned in 1958 at Shippingport in the United States and delivered 60 MW of electrical power at an overall efficiency of $\sim 26\%$. It utilised $\sim 3\%$ enriched UO_2 clad in zircaloy, an alloy of Zr with Sn, Fe and Ni. Zircaloy is widely used in water based reactor systems but is not suitable for gas-cooled reactor plants as it reacts with CO_2 to form carbides. Many light water reactors use stainless steel fuel cans, because of the superior strength, but this requires further fuel enrichment to $\sim 4\%$. The size and efficiency of PWR reactors have continued to improve and the newly built Sizewell B (UK) PWR station generates 1300 MW at an efficiency of $\sim 32\%$.

7.5.5 *Boiling Water Reactor*

The Boiling Water Reactor (BWR)[2] is an alternative to the PWR where the reactor core is designed to allow the coolant/moderator water to boil in the core (Fig. 7.15b). The potential advantage of such two-phase reactors is that the steam generated can be used to drive the turbines directly, eliminating losses in the intermediate heat exchangers. The system must, of course, be leak-tight to avoid release of potentially radioactive steam, although the BWR is comparatively safe, since any power surge leading to increased boiling in the fuel channels would result in a reduction of moderating power and a corresponding reduction in neutron thermalisation.

The first commercial scale BWR power station was built at Dresden in the USA and commissioned in 1960. The Dresden BWR was a light water system using 1.5% enriched UO_2 fuel clad in zircaloy and delivered 180 MW electrical power at an efficiency of $\sim 27\%$. The expected gains hoped for by elimination of the intermediate heat exchanger have not in practice materialised, and the BWR has not proved as successful a design as the PWR.

7.5.6 *RBMK reactor*

The principal Russian RBMK reactors, the first of which was installed in Leningrad in 1974, are pressurised, light water cooled, graphite moderated systems. The reactors use 2% enriched UO_2 fuel encased in zirconium-niobium alloy cans, which are arranged in vertical coolant channels surrounded by the graphite moderator. The reactors are typically arranged in pairs, each capable of driving two 500 MW turbines and therefore designated RBMK-1000.[g] Coolant enters the base of the reactor under pressure at $\sim 270°C$ in two loops, removing heat from the core as it

---

[g]Note: This is electrical power. The reactors generate considerably more thermal power — under normal operation the units produce 3200 MW of thermal power.

flows up through the channels forming wet steam (i.e. steam with a high moisture fraction). The wet steam is dried in separator units that remove the moisture component, which is returned to the main circulator pumps. The dried steam is used to drive the turbines, is condensed and also returned to the circulator pumps. Each loop is supplied by four circulation pumps, one of which is held in reserve under normal operation. In addition to Leningrad, additional RBMK-1000 reactors were installed at Kursk and Chernobyl; each based on four units built in pairs. The first of a larger RBMK-1500 series, designed to produce 50% more power, was installed at Ignalia in Lithuania in 1984.

7.5.7 *Advanced Light Water Reactor*

Concerns over safety, following the Three Mile Island incident (see Sec. 7.6) have led to the design of a new generation of light water reactors, the so-called Advanced Light Water Reactor (ALWR). A derivative of the PWR, the ALWR places a greater emphasis on *passive* safety features that minimise the likelihood of meltdown due to system failures (e.g. loss of coolant) and operator error. A quench pool positioned over the reactor housing would condense excess steam vented from the core in the event of a pressure build-up. If necessary, water in the quench pool could be released to flow down over the reactor. Most importantly, these passive safety features would come into play *irrespectively* of any operator or other normal control system intervention. The ALWR is also designed as a modular system, composed of separate units, each small enough to conduct sufficient heat from radioactive decay into the ground to prevent meltdown. With a construction time of about five years and a design life of 60 years, the ALWR is also intended to demonstrate to governments and the power generating utilities that nuclear power can be economic and competitive against conventional power generating systems.

The ALWR is proving to be the design of choice in South East Asia and France. Several are operational in Japan with others planned. South Korea also has a number of plants in operation with more in the construction or planning stage.

7.6 Safety of Nuclear Power

There have, in fact, been far fewer nuclear accidents than in other facilities of comparable size and complexity, e.g. chemical plants or oil refineries. However, two major incidents, Three Mile Island in the USA and Chernobyl in the Ukraine, have had a profound and possibly disproportionate effect on the public perception of the hazard posed by nuclear power stations. Proponents for nuclear power assert that such incidents could not happen again, but there is no such thing as a completely safe nuclear power station, just as there is no completely safe conventional power station. Prior to Three Mile Island there had only been minor incidents of little consequence or concern, but with the intense (and sometimes irresponsible) media interest, nuclear power stations became the focus of fear and mistrust.

The Three Mile Island[3,4] nuclear plant was a PWR system located near Harrisburg, Pennsylvania that suffered a partial meltdown due to a combination of design flaws and operator error. The incident occurred on 28th March 1979 when a lack of power to the feed-water pumps caused the steam generator to shut down automatically. However, the valve remained open allowing cooling water to drain out of the reactor. Since there were no sensors to indicate that the valve had failed to close, the operators misjudged the fault and reacted incorrectly to the emergency by shutting down the emergency cooling system. Inadequate instrumentation implied the reactor was full of water, when actually the reactor was in serious need of water. As a result, the core became uncovered for a time and suffered a partial meltdown as a result. Fortunately, the situation was brought under control and there were no injuries or deaths. The reactor was so severely damaged and contaminated that there are no plans to restart it and the clean-up costs inside the containment building have been substantial.

Chernobyl[5] has become the archetypical nuclear accident. Just about everything that could go wrong — flawed design, irresponsible operation, poor staff training — did. The accident occurred in the early hours of 26th April 1986 as engineers conducted a test to discover whether the turbine generator could supply some of the cooling pumps while slowing down to a standstill after the steam supply had been cut off. The experiment was carried out on the now infamous Unit No. 4 reactor to see if its power requirements could be sustained for a short time during a power failure. The experiment commenced in the early hours of April 25th when the reactor power was reduced to 50% of its normal operational level (1600 MW thermal). Later, at around 1.00 pm, one of the two turbines was tripped off the grid and all the power automatically transferred to the second unit, including four of the main coolant pumps. As part of the experiment, the reactor's emergency cooling system was disabled. At this point, there was a request from the grid controller in Kiev for the unit to continue supplying power which it did until about 11.00 pm when the experiment was resumed. During this period of ~9 hours the emergency cooling system remained disconnected, a violation of written operating procedures.

The experiment had been planned to take place with the reactor operating at between 700 MW and 1000 MW, but by midnight the power was already at the lower end of this range and continued to fall to 500 MW over the next 30 minutes. This triggered a change in the type of control rods from those used at high powers to a different set designed for low power operation. Unfortunately, the operators had failed to reset the operating point for these control rods and the power continued to fall rapidly to ~ 30 MW. The experiment should probably have been abandoned, nevertheless by removing many of the control rods, the operators were able to raise the power level to 200 MW — well below the target 700 MW. Later, analysis revealed that in doing so they had violated the minimum operating reactivity margin. The two standby by pumps (one in each loop) were switched on with the intention that at the end of the experiment, all eight pumps would be available to cool the reactor. The effect was a reduction in steam production and a concomitant loss of pressure in the separator units. The reactor should have tripped and shut down because of the low steam pressure in the separators, but the trip circuits had been overridden and pressure continued to decrease. All automatic control rods were withdrawn to maintain reactor power at 200 MW, and the power began to increase slowly. The reactor power then started to increase rapidly — becoming prompt critical (see Sec. 7.4.5.) — reaching 500 MW in $\sim$ 3s and continuing to increase exponentially. An emergency shut down was ordered, but since nearly all the control rods had by this time been removed there would have been a delay of about 20 s before reactor power could have been reduced. By then, the coolant water had boiled, leading to excessive fuel temperature and the onset of fuel channel rupture. Observers reported two large explosions which ignited fires that took several days to extinguish. Highly radioactive fission products were released into a plume that was carried on the prevailing winds as far west as Wales in the UK.[h]

Chernobyl occurred primarily as a result of human error and poor design. In particular, pump failure or cavitation (formation of bubbles) would augment the positive void coefficient, in itself a bit of dubious design. Any voiding could lead to a sudden local flash of steam with the possibility of fuel channel rupture, and local excess reactivity. That safety systems were disabled was a cardinal sin verging on the criminally negligent. When matters started to go wrong, the speed of events was such that the operators were unable to exercise any control. It also seems that there were no contingency plans in place to provide guidance for the operators. Subsequent enquiries highlighted a general lack of a safety culture among both operational and managerial staff — as manifest in the nine hours when the emergency coolant supply had been disconnected during the interruption to the experiment. One of the main conclusions was the need for a stronger regulatory regime able to counter pressure for energy production. In the event, two engineers were killed in the explosion and a further 31 personnel died shortly afterwards from radiation sickness. Estimates of additional worldwide cancer deaths vary from a few thousand to several hundred thousand.

7.7 Nuclear Waste

During the mid-20th century when the first technical developments enabled the deployment of nuclear power for civilian purposes, it was presumed that ways and means would be found to deal with the radioactive waste generated. There are two phases in the disposal of nuclear waste. Spent fuel rods that have been recently removed from the reactor core are too radioactive for safe handling and have to be

---

[h]Contamination of grass in certain parts of Wales have meant that sheep grazed on the affected fields were rendered unfit for human consumption and at the time of writing, they still are.

stored onsite in large water tanks. The water both absorbs the emitted radiation and cools the rods which will be physically hot as a result of radioactive decay. Over a period of ~ 10 years, the spent rods will lose over 95% of their radioactivity and may then be safely reprocessed; left over ^{235}U and ^{239}Pu which is also fissile can be removed for reuse and the remainder disposed of by geological burial, i.e. deep underground in stable rock formations.

Plutonium 239, is a product of the absorption by ^{238}U of a neutron forming the highly unstable isotope ^{239}U that rapidly decays by β emission to produce ^{239}Pu. The extraction of plutonium was seen to be an attractive option, effectively allowing use of the more abundant ^{238}U. However, plutonium is used for nuclear weaponry and consequently reprocessing of ^{238}U is banned in many countries, including the USA. Plutonium is in addition very toxic (quite far apart from being radioactive) and with a half-life of 24,000 years is one of the longest decay products. Given its weapons potential, the disposal of ^{239}Pu requires particular care.

It is generally agreed that geological burial is the only viable alternative for long-term containment. It was expected that radioactive waste 'dissolved' (by vitrification) in highly stable glass bricks would be stored in deep underground shafts in stable rock formations. Definitions of 'stable' vary from 10,000 to 1000,000 years, or in some cases, to a multiple of the longest half-life. The identification of a satisfactory location, from a purely geological point of view, is only part of the problem. Constructing the long-term containment facility invariably and understandably meets with considerable local opposition. It has so far proved impossible to allay fears that such facilities are safe and secure. To date, no geological burial has taken place, and both high and low level waste are still currently stored either onsite or at reprocessing facilities. As the volume of spent fuel increases, this mode of storage will become increasingly untenable, particularly in the USA, Japan and Europe where nuclear power has been in use the longest.

7.8 Problems

1. The fission of a ^{235}U atom produces atoms of ^{139}La and ^{95}Mo and 2 neutrons. Given that the atomic masses of ^{235}U, ^{139}La and ^{95}Mo are 235.0439u, 138.9061u and 94.9057u respectively, calculate the energy released. (1u = 931.161 MeV).

2. By comparing the collision of a neutron ($\sim$ unit mass) with a moderator atom (mass A) in the laboratory (L) frame of reference (where the moderator atom is considered to be stationary) with the same event in the centre-of-mass (CM) system of coordinates (where the centre-of-mass is stationary and the total linear momentum is zero), derive equation (7.6). [Hint: Sketch the situation in the two reference frames and using the condition that the total linear momentum in CM = 0, obtain expressions relating the velocities in L and CM.]

3. In a particular reactor, the fast fission factor, thermal utilisation factor and the mean number of neutrons produced per fission are 1.03, 0.89 and 1.34 respectively. If the fast and thermal leakage factors are 0.947 and 0.958

respectively, calculate the resonance escape probability when the reactor is just critical. If the resonance escape probability were to increase by 0.1%, determine the time it would take for the neutron flux to double.

4. Show that for very large reactors the critical equation becomes approximately:

$$k_{eff} \approx \frac{k_\infty}{1 + B^2(L^2 + \tau^2)}.$$

5. Show that for a parallelepipe-shaped reactor where the neutron density is

$$n(x, y, z) = n_0 \cos\left(\frac{\pi x}{a}\right) \cos\left(\frac{\pi y}{b}\right) \cos\left(\frac{\pi z}{c}\right)$$

where n_0 is the neutron density at the centre of the core, the geometric buckling must be given by equation (7.53).

References

1. http://www.dti.gov.uk/enegy/nuclear/technology/history/shtml
2. http://www.nrc.gov/reactors/power.htm
3. Crisis at Three Mile Island, *Washington Post Special Report* (1979) Online version available at http://www.washingtonpost.com/wp-sr/national/lomgtrm.tm/whathappened.htm
4. T. H. Moss and D. L. Sills, The Three Mile Island accident: Lessons and implications (New York Academy of Sciences, 1981). Proceedings of an academy meeting on the accident.
5. R. F. Mould, Chernobyl record, the definitive history of the Chernobyl catastrophe (IOP, 2000).

Chapter 8

RENEWABLE ENERGY

8.1 Introduction

Renewable energy sources are in fact not renewable![a] In reality what we mean by renewable are those sources of energy that are for all practical purposes inexhaustible, or else are rationed. It is also implied (often incorrectly) that they are non-polluting in use. In general, these are sources that in one way or another are supplied by the sun, either directly, as in the photovoltaic production of electricity and solar heating, or indirectly, as in wind energy and at one further remove, biomass energy. In a sense, fossil fuels are a stored form of solar energy; a result of photosynthesis in pre-historic times, and in burning them, we are simply releasing this energy. The problem of course is the time scale of replenishment.

Each year, the Earth receives $\sim 3.8 \times 10^{24}$ J of energy from the sun, far in excess of what we need. This energy is coming anyway and taking advantage of it will not change the global radiative balance. Harnessing that energy is the challenge. It is variable, distributed and unlike a gas fired power station, cannot always be simply switched on or off to meet fluctuations in demand. The grid based infrastructure established over the past several decades, and which has served so successfully to ensure continuity of supply, is largely incompatible (with the exception of large hydroelectric and geothermal facilities) with renewable energy sources. However, the case for diversification of energy supply away from fossil-fuelled generation is indisputable and the necessary resources will have to be deployed to achieve this.

In this chapter, we shall review the physics of some of the more important sources of renewable energy. In particular, we shall consider the direct use of solar energy to produce electrical power in photovoltaic solar cells and as a source of heat in water heaters and solar furnaces (sometimes misleadingly referred to as, respectively, active and passive uses of solar energy). We shall then consider indirect sources of solar energy, namely wind, biomass and hydroelectric. Finally, we shall discuss ground source heating and geothermal energy, a non-solar source.

---

[a] A "renewable energy source" would violate the laws of thermodynamics.

8.2 Photovoltaic Solar Cells

8.2.1 *The photovoltaic effect*

The *photovoltaic* effect may be defined as the generation of an e.m.f. through the absorption of a photon of light. It implies the generation *and* separation of negative and positive charge in much the same way as in a conventional battery. The separated charges may only recombine by the electron (i.e. negative charge) travelling round an external circuit and doing useful work on the way. Unlike the chemical battery, however, the photovoltaic solar cell cannot store useful quantities of power and effectively 'discharges' as soon as the illumination is switched off.

There are two essential requirements for the production of a photo-induced e.m.f.:

(i)   A material capable of absorbing the photon and producing the positive and negative charges (i.e. a semiconductor);

(ii) A means for separating the charges before they can recombine.

The latter requirement (ii) is fulfilled by diodes which we shall consider later. Before that we need to review some of the properties and characteristics of semiconductors and understand what makes them suitable for solar cell applications.

In crystalline semiconductors, electrons may only occupy bands of allowed energy levels separated by gaps of forbidden energy. In the present context, the important bands are the *valence* and *conduction* bands. At absolute zero ($T = 0\,\mathrm{K}$), the valence band is completely full of bound electrons while the conduction band is empty and contains no electrons. There are no *mobile* charge carriers and no current can flow. The two bands are separated by an energy or *band gap* (E_g) of an eV or two, and in their pure form semiconductors are resistive at low temperature. The injection of sufficient energy (i.e. in excess of the band gap energy), will excite a valence band electron across the gap into the conduction band where it becomes mobile. Additionally, the vacancy left in the valence band is also free to move, behaving as if it were a positive charge. The conductivity of the semiconductor is increased and it is able to conduct an electric current. The electron vacancies are referred to as *holes*.[b] Each electron excited to the conduction band leaves behind a hole in the valence band and there are consequently equal numbers of both types of charge carrier. In this case, the semiconductor is said to be *intrinsic*.

At any finite temperature ($T > 0\,\mathrm{K}$), there will be some electrons (however few) that have enough thermal energy to cross the energy gap and the distribution of

---

[b]There are many excellent texts on semiconductors and semiconductor devices, and the reader is referred to these for further information: e.g. S. Sze, *Physics and Technology of Semiconductor Devices*, John Wiley & Sons.

electrons between the two bands at any given temperature is given by the *Fermi-Dirac distribution function*.

$$F(E) = \frac{1}{1 + \exp((E - E_F)/k_B T)} \tag{8.1}$$

where E_F is the Fermi energy. The Fermi-Dirac expression gives the probability that some energy level E will be occupied by an electron at a given temperature T, and it is evident from (8.1) that for $(E < E_F)$, energy levels are likely to be populated and that the reverse is the case when $(E > E_F)$. For intrinsic semiconductors, the Fermi energy is located roughly midway between the top of the valence band and the bottom of the conduction band (Fig. 8.1).

The source of the energy is immaterial as long as it is larger than the band gap energy, E_g, and could be a *photon* of light, thus fulfilling the first requirement for the photovoltaic effect.

We can, in addition, control the relative numbers of electrons and holes through the judicious addition of impurities to the semiconductor, a process known as *doping*. Doping with an element that has a valency one greater than the host semiconductor (e.g. P (Group V) in Si (Group IV)) results in a 'spare' electron located on the impurity ion (Fig. 8.2a). The electron is loosely bound to the impurity and only a small amount of thermal energy is required to ionise the impurity, releasing the electron into the semiconductor as a mobile charge carrier, i.e. into the conduction band, in effect increasing the ratio of electrons to holes. The semiconductor is said to be *n-type*, the dopant is referred to as a *donor* as it has 'donated' an electron to the lattice, and the Fermi Energy is raised closer to the conduction band (Fig. 8.2b).

The opposite occurs if the dopant is an element with a valency of one less than the host (e.g. Al (Group III) in Si). The impurity atom is unable to satisfy all the bonds with the neighbouring semiconductor lattice, generating a hole in the valence band (Fig. 8.3a). There are thus more holes than electrons, the semiconductor is said to be p-type, the dopant is termed an *acceptor*, since it has 'accepted' an electron, and the Fermi Energy is lowered towards the valence band (Fig. 8.3b). In both n and p-type cases, the semiconductor is referred to as *extrinsic*.

The second requirement for generating a useful photo-induced e.m.f. can be met by creating some sort of potential barrier in the semiconductor. There are in essence two ways of doing this: at a metal semiconductor junction (Schottky diode); and in

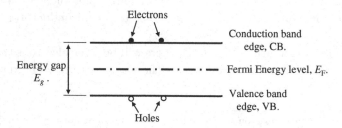

Fig. 8.1   Fermi level in an intrinsic semiconductor.

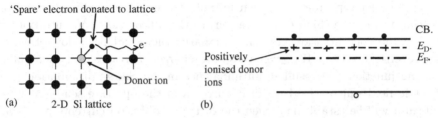

Fig. 8.2 Donor doping (a) schematic 2-D representation; (b) band diagram for n-type semiconductor.

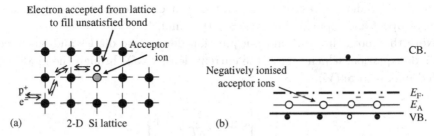

Fig. 8.3 Acceptor doping (a) schematic 2-D representation — unsatisfied bond (hole) exchanged with neighbouring atom; (b) band diagram for p-type semiconductor.

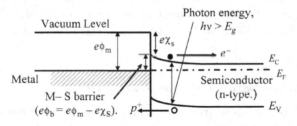

Fig. 8.4 Operation of a Schottky (M-S) barrier diode as a solar cell.

a p-n junction diode. The Schottky diode or metal-semiconductor (M-S) solar cell is illustrated in Fig. 8.4.

The mismatch between the metal work function ($e\phi_m$) and the semiconductor electron affinity ($e\chi_s$) results in a potential barrier ($e\phi_b$) at the semiconductor-metal interface. The potential barrier induces a space charge region in the semiconductor (represented in Fig. 8.4, by band-bending in the semiconductor), within which the electric fields are very high. Any electrons and holes excited into the high field region by photon absorption are rapidly swept apart, generating in the process an e.m.f. It was in this form that the photovoltaic effect was first reported in 1876 in a metal-Se device by Adams and Day, although it was not until more than half a century later that the device found widespread application as a photographic light meter.

Most solar cells are p-n junction devices, rather than M-S type structures. In principle, when a piece of n-type semiconductor is brought into intimate contact with a piece of p-type material, then initially there will be a diffusion current of holes from the p-type (high concentration) to the n-type (low concentration) sides driven by the concentration gradient. Similarly, there will be a flow of electrons in the reverse direction. However, these are charged particles and their diffusion across the junction will result in an imbalance in the charge distribution causing an electric field which will drive drift currents in the opposite sense. A dynamic equilibrium will be established when the drift and diffusion currents just balance. In equilibrium, the Fermi Energy must be constant throughout the device, which results in band-bending in the p and n sides of the junctions. The interface between the two semiconductors is consequently a region of high electric field, just as in the case of the M-S diode. Any photo-generated electron-hole pairs produced here will be swept apart and separated to generate the e.m.f. (Fig. 8.5).

Note that both the M-S and p-n junction devices are diodes and governed by the diode equation, which relates the current density (J) through the diode to the applied bias voltage (V).

$$J = J_s \left\{ \exp\left(\frac{eV}{k_\mathrm{B}T}\right) - 1 \right\}. \tag{8.2}$$

In forward bias ($V > 0$) the current density is dominated by the exponential term and current increases steeply with voltage. In reverse bias ($V < 0$), the exponential rapidly vanishes and the current density saturates at $-J_s$, which is why it is called the *Reverse Saturation Current Density* and is a property of that particular diode.

8.2.2 *Performance parameters for solar cells*

Solar cells are a source of electrical power, i.e. the product of current and voltage, and the principal characteristic we need to know is the voltage and current that can be delivered under a given level of illumination; the *Photovoltaic Output Characteristic*, a graph of current against voltage. It takes the same form as a

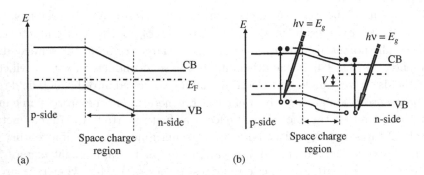

Fig. 8.5   Band diagrams for a p-n junction (a) in equilibrium; (b) under illumination.

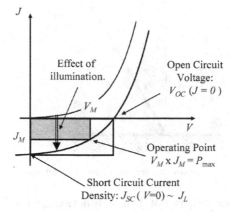

Fig. 8.6   Solar cell photovoltaic output characteristics.

diode characteristic, except that it is shifted down the current axis by an amount equal to the light-induced current density,[c] J_L, as shown in Fig. 8.6.

The important parameters are the *open circuit voltage* (V_{oc}), *short circuit current density* (J_{sc}) and the *fill-factor* (*FF*). The open circuit voltage is the voltage we would measure across the output terminals of an illuminated solar cell when no load is connected ($J = 0$). Similarly, the short circuit current is the maximum current that the cell can provide (i.e. when the terminals are short circuited; $V = 0$). Although V_{oc} and J_{sc} are useful as figures of merit, what we really need to know is how much power we can extract from the cell, and this is given by the product $(V \times J)$ W m^{-2}. Under open circuit conditions, V is a maximum but $J = 0$ and so $V.J = 0$, similarly, in short circuit the current is a maximum but V is zero. Between these two extremes there will be some optimum or *operating point* (V_m, J_m) when the power is a maximum. The fill-factor is the ratio of the product of $V_m J_m$ to $V_{oc}J_{sc}$ .It corresponds to the fraction of the output characteristic that is actually useful, hence the name.

The *conversion efficiency* η is simply the ratio of the output power at the operating point to the incident radiant power density, P_0 (W m^{-2}).

$$\eta = \frac{V_m J_m}{P_0} = \frac{V_{oc}J_{sc}FF}{P_0}. \tag{8.3}$$

For purposes of inter-comparison between cells we define standard atmospheres in terms of the corresponding insolation at some zenith angle (z_s) on a clear day at sea level. The path length through the atmosphere is directly proportional to $\sec(z_s)$; the greater the zenith angle the longer the path length and hence the mass of air the light has to traverse (Fig. 8.7). We use this concept of air mass to define standard levels of illumination. For example, air mass 1 (AM1) is when the sun is directly overhead, $z_s = 0^0$, $\sec(z_s) = 1$.

---

[c]The photo-generated electron-hole pairs are separated in the opposite sense to the diffusion current and therefore increase the reverse-bias current.

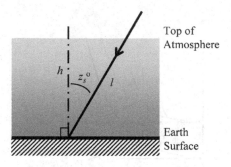

Fig. 8.7   Diagram illustrating concept of air mass number.

The insolation, however, is not just a function of the air mass, because the absorption of light in the atmosphere is neither spectrally neutral nor uniform with altitude, and air mass number is only approximately equal to $(1/\cos(z_s))$. A widely used standard of comparison is 'AM1.5' which corresponds to a total irradiance of $1\,\mathrm{kW\,m^{-2}}$ (See Fig. 3.1) with a defined sea level spectrum.

The choice of a suitable semiconductor is an interesting one and illustrates the common experience in practical situations of having to balance advantages and disadvantages. We might imagine that we would need a semiconductor with a small band gap energy to capture as much of the incident radiation as possible, (i.e. $E_g \ll h\nu$). Each absorbed photon would contribute one electron-hole pair to the electric current which in consequence would be high. However, while it is true that solar cells fabricated from narrow gap semiconducting materials do absorb a very high proportion of the incident solar radiation and convert it to electric current, they do so at rather low output voltage. As we have seen above, the requirement from a solar cell is power (i.e. JV), and if V is low then output power will be low too. We would obtain a higher output voltage from a semiconductor with a larger E_g, but at the cost of absorbing less of the incident solar radiation and the current and hence the power would still be low. There is consequently a trade-off to be made between current and voltage output giving an optimum band gap energy of $\sim 1.3\,\mathrm{eV}$.

The operation of an ideal solar cell may be described in terms of a simple equivalent circuit consisting of an ideal current source[d] representing the light-induced current in parallel with a diode as shown in Fig. 8.8(a). The light-induced current passes through the diode generating a potential V_D across it. In the ideal case, the diode voltage is equal to the output voltage, V_0. Equating currents in the circuit, we can write down the output current density J_0:

$$J_0 = J_s \left\{ \exp\left(\frac{eV_D}{k_{\mathrm{B}}T}\right) - 1 \right\} - J_L \approx J_s \exp\left(\frac{eV_D}{k_{\mathrm{B}}T}\right) - J_L \qquad (8.4)$$

from which we find that since $J_S \ll J_L$, the short circuit current is:

$$J_{sc} = -J_L : \quad \text{when } V_0 = 0 \qquad (8.5)$$

---

[d]An ideal current source is one that can supply a constant current irrespective of load.

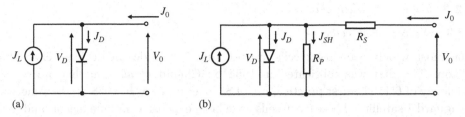

Fig. 8.8 Equivalent circuits for (a) ideal solar cell; (b) real cell with series and shunt losses.

and the open circuit voltage is:

$$V_{oc} \approx \frac{k_B T}{e} \ln \left\{ \frac{J_L}{J_s} \right\} \quad \text{when } J_0 = 0. \tag{8.6}$$

Real solar cells are not ideal and suffer from internal losses that act to degrade the performance. We categorise these losses as *series* losses (e.g. resistance of the semiconductor, contacts etc.) which we represent in the equivalent circuit (Fig. 8.8b) by a resistance in series (R_s) with the output, and *shunt* or *leakage* losses which we represent as a resistance (R_p) in parallel with the diode. It is evident that the output voltage is no longer equal to the diode voltage as some potential is dropped across the series resistance. Similarly, the output current is reduced by the leakage current flowing through R_p. Generally, R_s and R_p are normalised by area, i.e. $\Omega \, \text{m}^2$ to provide a better means of comparison.

$$J_0 = J_s \left\{ \exp\left(\frac{eV_D}{k_B T} \right) - 1 \right\} - \frac{V_D}{R_p} - J_L : \quad V_D = V_0 - J_0 R_s. \tag{8.7}$$

The short circuit current is reduced because when we set V_0 to zero, $V_D = J_0 R_s \neq 0$, hence:

$$J_{sc} \equiv J_0 \approx J_s \exp\left(-\frac{e J_0 R_s}{k_B T} \right) + \frac{J_0 R_s}{R_p} - J_L \approx \frac{J_0 R_s}{R_p} - J_L. \tag{8.8}$$

In other words, J_{sc} is reduced by the term in R_s and it is clear from (8.8) that for a high short circuit current we require R_s to be small. Internal losses also reduce the open circuit voltage. Setting $J_0 = 0$ in (8.7):

$$J_L \approx J_s \exp\left(\frac{eV_{oc}}{k_B T} \right) - \frac{V_{oc}}{R_p} \approx \quad \text{or} \quad V_{oc} \approx \frac{k_B T}{e J_s} \ln \left\{ \frac{J_L}{J_s} \right\} \quad \text{if } R_p \text{ is large.} \tag{8.9}$$

Equation (8.9) shows that we need a large value of R_p, in which case the exponential term becomes dominant and the equation approximates the ideal case, (8.6). Series and shunt resistance losses also reduce the fill factor, decreasing efficiency even further.

8.2.3 *Types of solar cell*

8.2.3.1 *Silicon solar cells*

The most widely used solar cells are p-n junction diodes made from crystalline silicon. The first was fabricated in 1954 by Chapin *et al.*[1] and had an overall efficiency of 6%. The first practical use of Si solar cells was in 1958 on the American Vanguard 1 satellite. These early cells were both expensive and inefficient, and until the oil crises of the 1970's, the utilisation of solar cells was confined to applications where either cost was secondary to other issues (e.g. in space) or where owing to remoteness, there was no alternative. Continuing development and the increasing cost and instability of oil supplies due to international politics, coupled with the growing concern about global warming has led to the increased use of photovoltaics, and they are now recognised as an important source of renewable energy.

There are in principle three types of silicon-based solar cell, defined by the crystallinity:

(i)   Mono-crystalline cells are the most efficient but also the most expensive;

(ii) Multi-crystalline devices are less efficient, but considerably cheaper to produce with much less wastage;

(iii) Amorphous thin film Si cells are the least efficient but also the least expensive. We shall discuss the amorphous Si cells later along with other thin film types (Sec. 8.2.3.2).

Single crystal or mono-crystalline Si cells are at the high performance end of the solar cell device spectrum, in spite of a band gap energy (1.1 eV) well below the optimum and poor optical absorption characteristics. However, this is compensated by the high level of Si technology, which allows the fabrication of complex device structures that mitigate the disadvantages. The cells are formed on large diameter n-type wafers into which thin p-type regions have been diffused to form the p-n junction (Fig. 8.9). The surface is usually strongly textured, typically into an inverted pyramid-like pattern, in order to trap the incident radiation, and coated with an anti-reflection coating to minimise reflection losses. Advanced surface passivation and contact techniques reduce losses further. As a result, Si cells produce high currents, although at modest output voltages. Nevertheless, single cells typically can deliver efficiencies of over $\sim$24%.[2]

The structure of multi-crystalline silicon solar cells is not greatly different from those of single crystal devices. However, here the Si is composed of many individual crystal grains, with additional losses at the grain boundaries. Conversion efficiencies are reduced by $\sim$4% compared with the single crystal devices.

Solar cells are seldom used as individual single cells, and for power generation many individual cells have to be connected in various parallel and series configurations — depending on the application and circumstances — to form modules. This introduces further losses from interconnection resistances and module aperture, which arises from edge and shadow effects. Single crystal ingots of Si are

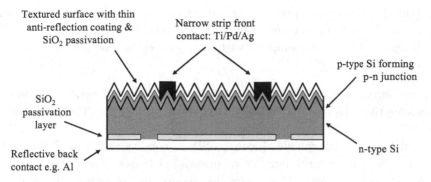

Fig. 8.9   Schematic diagram of Si solar cell.

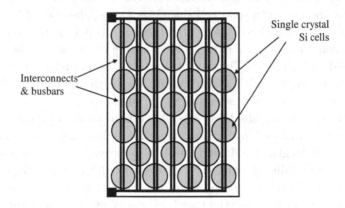

Fig. 8.10   Mono-crystalline Si solar cell module illustrating aperture loss.

produced by pulling a seed from a crucible of molten Si, and are roughly cylindrical in shape. The ingots are then diced to form circular wafers into which the cells are fabricated as described. It is difficult to assemble circular discs into a module without wasting a lot of either the available space (Fig. 8.10) or material.

Multigrain Si, on the other hand, can be produced as a continuous (rectangular) ribbon. Module coverage is therefore much better, and the difference in performance between single and multigrain Si modules is in practice not so great.[2]

8.2.3.2 *Thin film solar cells*

Thin film ($< 10\,\mu$m in thickness) structures are an alternative approach to the very complex, relatively expensive Si solar cells. Thin film solar cells are predicated on three ideas:

(i)   Low usage of expensive semiconducting and potentially polluting materials (while Si itself is not polluting, many of the chemicals used in the production of Si cells are highly toxic and polluting);

(ii)   Low energy consumption in manufacture (unlike Si, there is no energy
       intensive component comparable to the growth of the initial crystal ingot);
(iii) Integrated modules, complete with interconnects can be produced in a single
       production run.

There are three main contenders for thin film solar cells: amorphous silicon (α-Si),
Cu-chalcopyrite cells (mainly $CuInSe_2/CdS$ and $Cu(In,Ga)Se_2/CdS$) and the CdS/
CdTe cell.

By definition, amorphous silicon (α-Si) is not crystalline and the material
is characterised by a large number of unsatisfied bonds. These generate a high
concentration of electron and hole traps that reduce the carrier density and mobility.
The incorporation of hydrogen passivates the incomplete bonds, greatly increasing
the mobility of electrons and holes in the material. It can then be doped to create
a p-n junction and used as a solar cell since it is a good absorber of light, although
the band gap is a little high at 1.75 eV. In practice, it is better to insert an intrinsic
(i) layer between the p and n regions to form a p-i-n diode, where the electric field
is concentrated in the i-layer. The device (Fig. 8.11) is normally produced on glass
that has been pre-coated with a transparent conducting oxide (e.g. tin oxide or
indium tin oxide (ITO)) which serves as the front contact.

The α-Si film is deposited by plasma decomposition of silane (SiH_4) and
the structure is completed by a back metallic highly reflective contact of Al.
Unfortunately, hydrogen passivated α-Si has proved to be unstable under
illumination — something of a handicap for a solar cell! Consequently, α-Si cells are
unlikely to be used in large scale power systems, but they have found widespread
application in low power applications such as solar-powered calculators.

The Cu-chalcogenide and CdTe-based thin film cells are heterojunctions, that
is p-n junctions created from two different semiconductor materials. They are based
on the 'window-absorber' model (Fig. 8.12). Both $CuInSe_2$ and CdTe are such
strongly absorbing semiconductors that a film just a few microns thick is totally
opaque. Consequently, it is difficult to make a photovoltaically useful p-n junction
in either since most of the incident light would be absorbed before reaching the
active part of the device. In a window-absorber structure, a different semiconductor
with a much wider energy gap and therefore for practical purposes transparent is

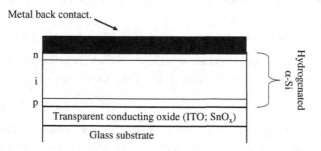

Fig. 8.11   Schematic diagram of α-Si solar cell.

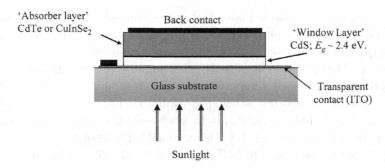

'Absorber layer'
CdTe or CuInSe$_2$

Back contact

'Window Layer'
CdS; $E_g \sim 2.4$ eV.

Glass substrate

Transparent
contact (ITO)

Sunlight

Fig. 8.12   Schematic diagram of CdS/CdTe window-absorber cell.

used as the n-side limb of the p-n junction. Incident light passes through the wide gap semiconductor (window) layer to the junction where absorption and charge separation are efficient.

Copper indium diselenide has a band gap energy of about 1.1 eV, which is too small so it is often alloyed with a small fraction (x) of Ga (CuIn$_{1-x}$Ga$_x$Se$_2$) which increases the band gap. It is usually paired with CdS ($E_g = 2.45$ cV) which acts as the transparent window. Cell efficiencies up to 18.8% have been reported.[2] The n-CdS/p-CdTe thin film cell is potentially the most efficient with a theoretical maximum efficiency of $\sim 28\%$, although the highest value reported to date is 16.8%.[2] Small modules (0.1 m^2) are rather less efficient than this and typical conversion efficiencies are in the order of 7–10%.

8.2.3.3 *Utilisation of photovoltaic solar power*

Crystalline Si cell technology is well developed, reliable and mature. Incremental improvements may reasonably be expected to increase module efficiency and reduce cost. Multi-crystalline modules are beginning to permeate the energy market providing local sources of energy for low power applications such as parking meters, domestic water pumps and remote weather stations. Thin film technologies are much less well-developed and have yet to find widespread use. Little is known about their long term operational characteristics and degradation mechanisms.[3]

In the generation of photovoltaic power, expenditure scales directly with the area. Module costs, land requirements and supporting infrastructure are all area-dependent and combine to make large installations very expensive. Clearly, high module efficiency/cost ratios are required to make photovoltaic power generation competitive with other sources. Although capital costs are high, recurrent costs are relatively low and so the total cost for power varies inversely with the lifetime of the plant. Silicon modules have lifetimes of about twenty-five years, but are expensive. It is generally agreed that thin film modules need to achieve efficiencies above about 10% and match the 25 year lifetime of the Si modules if they are to find large scale application.

The sizing of photovoltaic power systems is critically important. If oversized, the price of the power supplied becomes too high as capital expenditure is such a major component of the cost. If undersized, the station will prove to be unreliable in operation. The question is of particular importance in temperate climates, where there is a substantial difference between winter and summer insolation and moreover, demand is greatest when the available supply is at a minimum. In reality, annual and diurnal variations in photovoltaic supply require integration with the main electrical grid system. This is an added level of complexity (and cost) requiring the use of inverters (DC/AC converters), and although inverter conversion efficiencies are high, there are still additional energy losses to contend with.

The annual rate of production of photovoltaic modules is expected to reach several million square metres over the next decade. This will pose major issues for cells containing scarce materials, principally the copper chalcogenides — both In and Ga are rare metals and are widely used in competitor industries — microelectronics, optoelectronics, flat panel displays etc. Although Cd is a more abundant element than both Ga and In and is not subject to the same competitive pressures, it is more toxic. The environmental implications of using CdTe cells are consequently more severe and there will need to be strict controls on both production and disposal. The production of all types of solar cell involve some degree of pollution in terms of the reagents and processes consumed and to the extent that heating and transport is involved, the emission of CO_2. These issues are not always factored into the environmental costs.

8.3 Thermal Solar Power

8.3.1 *Solar collectors*

If the level of insolation is sufficiently high, then solar energy may be used to provide hot water for domestic uses. The technology need not be sophisticated and is ideally suited to relatively underdeveloped communities. At its simplest, water is passed down tubes embedded in a solar collector (Fig. 8.13). Absorption of heat from the incident solar radiation raises the temperature of the collector to ∼80°C, and this heat is then transferred to the water flowing through the pipes.

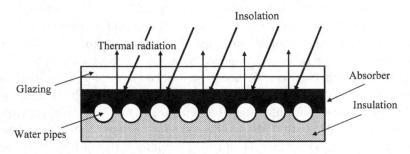

Fig. 8.13   Solar water heater.

Although conceptually simple, there are some practical issues that must be taken into account. In order to maximise the heat gain by the collector, it should be painted black so that it has a high absorption coefficient ($\alpha \sim 1$) for the incoming solar radiation. However, at $\sim 80°C$, radiative loss[e] to the surroundings would be from (5.2) $\sim$460 Wm^{-2}. The net gain is modest, when compared with the incoming radiative power (~ 700 Wm^{-2}), especially given that there will be convective and conductive losses as well. It is important, therefore, that the collector coating have a high absorption coefficient at the peak solar wavelength ($\lambda = 500$ nm) and a low emissivity (ε) at the corresponding emission wavelength $\lambda \sim 8\mu$m; i.e. $\alpha(500$ nm$) \gg \varepsilon(8\mu$m$)$. The net heat supplied to the water in the pipes, may be estimated by calculating the energy balance in steady state. For unit area of collector:

$$q_i = t_g \alpha S - \sigma\varepsilon(T_{Col}^4 - T_{Air}^4) - h(T_g - T_{Air}) \qquad (8.10)$$

where t_g, S, σ, h, T_{Col}, T_{Air} and T_g are the transmissivity of the glazing, the solar insolation, Stephan-Boltzman constant, coefficient of convection for the glazing, and the collector, air and glazing temperatures respectively. (We have ignored conductive losses to the environment on the assumption that these can be made small by proper insulation). In steady state q_i must equal the heat being removed by the water. If m kg s^{-1} of water is flowing through the collector then the increase in water temperature is:

$$\Delta T_{\text{H}_2\text{O}} = \frac{q_i}{mc_p} \qquad (8.11)$$

where c_p is the specific heat at constant pressure of water. In well-designed systems, temperature increases of $\sim$25°C can be achieved, quite adequate for domestic purposes. The heated water must then be stored in a well-insulated tank until it can be used, if solar heated water is to be effective.

8.3.2 *Generation of electricity by solar heat*

Solar radiation may also be used as a source of heat in the production of electricity. Efficient generation requires high temperatures (Sec. 6.2.3) which may be achieved using mirrors to concentrate the solar radiation in a furnace. Several configurations have been investigated using elliptical and spherical mirrors to focus the radiation onto pipes carrying working fluids. These act as heat pipes conveying the energy to efficient externally heated engines such as the Stirling engine which, using a conventional crank, turns an electrical generator.

The Stirling engine is a reciprocating engine consisting of two constant volume processes and two isothermal processes (Fig. 8.14a). The operation of the engine is best understood in terms of the $T - S$ diagram (Fig. 8.14b). The working fluid

---

[e]By Kirchhoff's Law, emissivity (ε) = absorption coefficient (α) at a given wavelength.

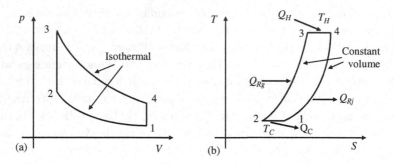

Fig. 8.14   Stirling engine : (a) $p - V$ diagram; (b) $T - S$ diagram.

(a gas) is first compressed (1→2) isothermally, releasing heat Q_C in the process. Heat, Q_{Rg} is supplied at constant volume (2 → 3), raising the temperature from T_C to T_H. The gas is then expanded isothermally (3 → 4) (which can only be done by the supply of an appropriate amount of heat, Q_H) doing useful work, W in the process. Finally, heat Q_{Rj} is rejected at constant volume (4 → 1) and the temperature is returned to T_C to restart the cycle.

So long as the specific heat at constant volume remains constant over the cycle, then the amount of heat required to raise the temperature of the gas from T_C to T_H at constant volume will be the same as the heat released on cooling the *same mass* of gas at constant volume between the two temperatures; $Q_{Rj} = Q_{Rg}$. More importantly, it is technically possible to 'recycle' most (> 90%) of the rejected heat in a regenerator significantly improving the efficiency:

$$\eta = \frac{W}{Q_H} \approx \frac{Q_H - Q_C}{Q_H} = 1 + \frac{Q_C}{Q_H}. \tag{8.12}$$

This is approximately equal to the Carnot efficiency (6.17). If the rejected heat, Q_{Rj}, is reused then it is not included in the calculation of efficiency (8.12), since to do so would be to count the same energy twice.

The regenerator (Fig. 8.15) is composed of a matrix of small pipes, across which a temperature gradient is maintained between the two isothermal stages (1–2 and 3–4). Fluid enters the matrix in state 4, transfers heat to the matrix and leaves at state 1. The working fluid in state 2 flows through the matrix in the reverse direction, gaining heat and leaving at state 3. Since the working fluid and matrix

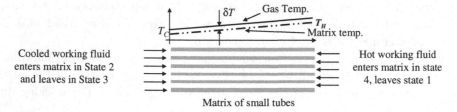

Fig. 8.15   Stirling energy regenerator.

only differ infinitesimally in temperature at each point, the process is effectively reversible, and the efficiency of the cycle is close to the Carnot efficiency.

Currently, there are plans to build a very large (500 MW) solar thermal station based on Stirling Engines in the Mojave Desert of Southern California north-east of Los Angeles. The station will use an array of 20,000 dishes to focus sunlight onto the engines. The Mojave facility will augment conventional power sources during daylight hours when the soaring use of air conditioners in the afternoon leads to a peak in demand. The station will generate 'grid ready' A.C. peak power at an efficiency of $\sim 29\%$, roughly twice that of photovoltaic sources.

8.4 Wind Power

8.4.1 *Principles of wind power*

The harnessing of wind energy is nothing new. It has been used in sailing ships since biblical times and was used in the medieval period to power wind mills for grinding corn. With the advent of the industrial revolution in the early nineteenth century, wind energy was superseded by steam systems and then in the 20th century by oil and gas as the principal motive force. Sailing became a leisure activity, and wind mills a picturesque component of the countryside. Things have changed in recent decades; 'wind mills' in their modern guise as wind turbines are becoming a relatively common, if sometimes controversial feature of the rural skyline, as a carbon-free source of power.

Energy is extracted from the wind through the differential flow of air over an aerofoil (Fig. 8.16), which produces a pressure difference that in the case of sailing ships, provides the motion, or in a turbine, the rotation. (The same principle is used in aircraft to provide lift, although here the 'wind' is artificially created by the speed of the plane moving through effectively stationary air.) The blade of a modern wind turbine is designed so that air flowing over one side must travel a larger distance than on the reverse side (Fig. 8.16a). Air flowing over the top (the longer path) then speeds up, while that flowing underneath slows down. According

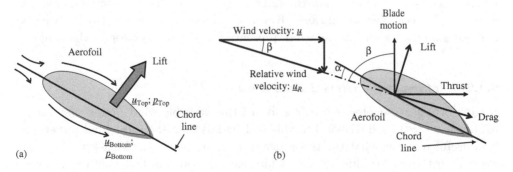

Fig. 8.16   Air flow over aerofoil.

to Bernoulli's law:

$$p + \frac{1}{2}\rho u^2 = \text{constant} \tag{8.13a}$$

where p, u and ρ are the pressure, wind speed and air density respectively. Since the velocities are not equal, it follows there will be a net pressure difference which provides the lift:

$$F_{\text{lift}} = p_{\text{bottom}} - p_{\text{top}} = \frac{1}{2}\rho(u_{\text{top}} - u_{\text{bottom}}). \tag{8.13b}$$

Strictly, Bernoulli's law applies to incompressible, non-viscous fluids, which clearly air is not, but (8.13b) does at least serve as a qualitative explanation for the lift forces.

For proper functioning, the chord line of the aerofoil should make the correct angle (known as the angle of attack) with the relative wind direction (Fig. 8.16b); that is the resultant wind direction between the wind velocity vector and the turbine blade (tangential) velocity vector. The latter is a function of position along the blade; tangential velocity is considerably greater at the turbine tip (especially in large diameter systems) than near the hub. Consequently, to maintain optimal performance, the blades are sometimes twisted along their length — although since the wind speed is not a constant this can only be of limited advantage. In addition to the lift, there will be a drag force due to the loss of momentum of the wind which must be absorbed by the tower. Wind turbine blades are designed to provide lift forces 20 to 30 times greater than the drag forces.

If the wind speed is u, then in unit time the volume of air flowing through unit area is u m^3, and as the kinetic energy of the wind will be $1/2\rho u^2$, then the total wind energy that passes through unit area in one second is:

$$P_w = \frac{1}{2}\rho u^2 \times u = \frac{1}{2}\rho u^3. \tag{8.14}$$

The cubic dependence of power on the wind speed represents both an advantage and a problem. To maximise the output power, we should site wind turbines at locations where the mean wind speed is high, and clearly a site with even a small increase in mean wind speed offers significant advantages. However, the cubic dependence also magnifies any variations in the wind, presenting a serious engineering challenge.

8.4.2 *Performance criteria: Betz limit*

An important parameter is the fraction of the available wind power that can *in principle* be extracted. Inevitably, this will be less than the total wind power, but it is useful to know what the theoretical maximum power is. We assume that the mass of air that passes through the turbine can be treated as being entirely separate from the surrounding air. We represent the affected air mass as being confined in a

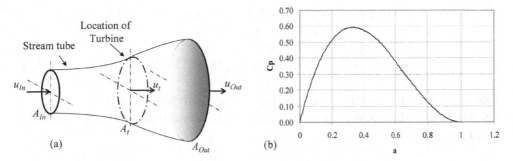

Fig. 8.17 Calculation of Betz limit (a) geometry; (b) variation of C_P with a.

'stream tube' of circular cross-section (Fig. 8.17a) that includes at some point the turbine, i.e. at the turbine, the diameter of the stream tube is equal to that of the turbine, A_t.[4] By definition, the mass flow rate must be a constant through every cross-section of the tube, in other words mass continuity applies.

The air entering the tube far upstream of the turbine will be slowed by the presence of the turbine, and in consequence the cross-sectional area of the tube will increase in order to accommodate the requirement for mass continuity. As the air passes through the turbine, there is a drop in pressure so that on the down stream side — known as the wake — the pressure is below that of the surrounding atmosphere. Far downstream in the wake equilibrium will be re-established and the pressure will be restored to that of the surroundings. If the unperturbed wind velocity entering the stream tube is u_{in}, and the tube area at this point is A_{in}, the mass flow rate entering the tube is $\rho A_{in} u_{in}$, where ρ is the density, and this must be the same at all points along the tube, in particular at the turbine and in the wake:

$$\rho A_{in} u_{in} = \rho A_t u_t = \rho A_{out} u_{out} \tag{8.15}$$

where A_t is the area swept out by the turbine blades, A_{out} the cross-sectional area of the wake and u_t and u_{out} the wind velocities through the turbine and in the wake.

It is customary to characterise u_t in terms of the free stream wind speed (u_{in}) less a subtracted component $a u_{in}$ such that:

$$\frac{u_t}{u_{in}} = (1 - a) \tag{8.16}$$

where a is variously referred to as the axial flow induction factor, the axial flow interference factor or simply the inflow factor. As the wind passes through the tube there will be a change in momentum due to the overall change in the wind velocity (mass flow rate is constant). There is also a drop in static pressure across the turbine, and the product of this with the area swept out by the turbine blades provides the force producing the loss of momentum.

$$(u_{in} - u_{out})A_t\rho u_t = (u_{in} - u_{out})A_t\rho u_{in}(1-a) = A_t(p_+ - p_-) \qquad (8.17)$$

where p_+ and p_- are the pressures just before and just after the turbine. The upstream and downstream pressures may be obtained from Bernoulli's equation (8.13a)[f]:

$$p_{in} + \frac{1}{2}\rho u_{in}^2 = p_+ - \frac{1}{2}\rho u_t^2 \quad \text{and} \quad p_{out} + \frac{1}{2}\rho u_{out}^2 = p_- + \frac{1}{2}\rho u_t^2. \qquad (8.18a)$$

Subtracting the two equations and recalling that far downstream, where equilibrium has been restored, the pressure in the wake is the same as that far upstream, $p_{in} = p_{out}$:

$$(p_+ - p_-) = \frac{1}{2}\rho(u_{in}^2 - u_{out}^2). \qquad (8.18b)$$

Substituting in (8.17) we get:

$$(u_{in} - u_{out})A_t\rho u_{in}(1-a) = \frac{1}{2}\rho A_t(u_{in}^2 - u_{out}^2). \qquad (8.19a)$$

From which we get:

$$u_{out} = (1-2a)u_{in}. \qquad (8.19b)$$

The power extracted from the wind by the turbine is given by the product of the force (8.17) and the wind velocity, u_t:

$$P = \frac{1}{2}A_t\rho u_t(u_{in}^2 - u_{out}^2) = 2A_t\rho u_{in}^3 a(1-a)^2. \qquad (8.20)$$

Apart from the inflow factor, a, the last term in (8.20) is expressed in quantities that can be measured experimentally.

We define a *Coefficient of Performance*, C_p, as the ratio of the maximum power that can be extracted from the wind to that available, P_w (8.14):

$$C_p = \frac{P}{P_w} = \frac{2A_t\rho u_{in}^3 a(1-a)}{(1/2)A_t\rho u_{in}^3} = 4a(1-a)^2. \qquad (8.21)$$

---

[f]Bernoulli's equation states that under steady conditions the total energy (i.e. static pressure energy + kinetic energy) in the flow is a constant; here total upstream energy $\neq$ downstream energy, so we must treat each side separately.

The denominator in (8.21) is the power in the wind without the perturbing influence of the turbine.

To find the optimum coefficient of performance we need to differentiate P with respect to a and equate to zero in the usual way:

$$\frac{dC_p}{da} = 4\{(1-a)(1-3a)\} = 0; \quad \text{i.e. } a = \frac{1}{3} \quad \text{and} \quad C_p = \frac{16}{27}. \tag{8.22}$$

This is known as the *Betz Limit* after the German aerodynamicist. It is the absolute upper limit of performance for a wind turbine (Fig. 8.17b). The limit arises because the stream tube has to expand upstream of the turbine as a consequence of the back pressure experienced by the approaching air. It has nothing to do with the design of the turbine.

8.4.3 *Vertical dependence of wind speed*

An implicit assumption in the above discussion is that the wind velocity is homogenous and essentially constant. This is never the case and there is a relatively strong dependence of wind speed with height above ground (Fig. 8.18).

Frictional forces progressively reduce wind speed as altitude is decreased, until at the surface the air is essentially stationary in a boundary layer, dominated by frictional forces. The thickness of the boundary layer, δ, may be estimated from the Reynolds number, Re, a dimensionless quantity defined by the ratio of the inertial and viscous forces. For a fluid of density ρ:

$$Re = \frac{\rho u_* \delta}{\eta^*} \tag{8.23}$$

where u_* is a characteristic velocity sometimes referred to as a shear or friction velocity and η^* is the shear viscosity. These are empirical values to be determined

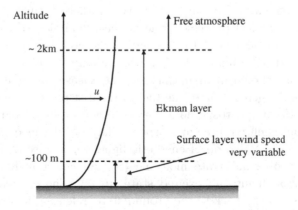

Fig. 8.18   Variation in wind speed with altitude.

by measurement in any given circumstance. Within the boundary layer, frictionless forces are dominant implying a low $Re \sim 10$. Above the surface layer, wind speed increases asymptotically to its unperturbed value. To a first approximation, we may assume that the rate of change of u with height, z decreases inversely with z:

$$\frac{\partial u}{\partial z} = \frac{u_*}{kz}: \quad \text{or} \quad (z) = u_* \ln\left(\frac{z}{z_0}\right) \tag{8.24}$$

where z_0 is a measure of the surface roughness and depends on the nature of the surface coverage, e.g. grass, trees, etc. Although (8.24) holds well up to a height of several hundred metres, it implies $u(z_0) = 0$, and is only an approximate description at z close to z_0.

Equation (8.24) allows us to relate the wind speed at the wind turbine, perhaps a few tens of metres above ground, with the geostrophic winds at several hundred metres. It is the geostrophic winds that can be predicted with greater precision in weather forecasts and therefore allow optimum operation of the turbine as an element within the power supply grid.

There have been significant advances in wind turbine technology in recent years, and large turbines ($> 2\,\text{MW}$) are presently being installed throughout Europe and the USA, in both onshore and off-shore locations. The latter offer significant advantages in that there is less visual impact and winds are less variable on short time scales, although they can reach significantly higher levels in storm conditions, placing tower and infrastructural components (not to mention the blades) under considerable strain. Offshore sites are also more hostile environments, making maintenance and repair more difficult and turbine components will be subject to severe corrosion from sea water.

8.5 Biomass Energy

8.5.1 *Photosynthesis*

The term 'biomass energy' really covers two rather distinct types of energy production from biological processes: one is the production of carbohydrates by photosynthesis for burning and the second is the creation of biogas (primarily methane) by the action of anaerobic (i.e. in the absence of air) bacteria on waste. Both processes are carbon neutral, in that the CO_2 released by combustion is merely that consumed in the generation of the fuel in the first place.

The processes of photosynthesis are rather complex, but a qualitative understanding sufficient for present purposes may be obtained from the energy diagram in Fig. 8.19. A photon of energetic light, $h\nu_i$ is absorbed by the ground state combination of a substrate molecule (S) and chlorophyll molecules (Ch), raising the chlorophyll into an excited state, (Ch*). The excited Ch* molecule may spontaneously decay back to the ground state, releasing its energy essentially as heat, or alternatively it may convert the substrate molecule S into a 'product'

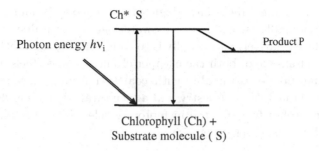

Fig. 8.19 Simplified energy diagram for photosynthesis.

molecule, P, providing chemical storage of the energy. In this respect, photosynthesis differs from photovoltaic and wind sources which do not have energy storage capability.

The net chemical reaction describing the production of carbohydrates by photosynthesis is:

$$6H_2O + 6CO_2 + 4.66 \times 10^{-18}J \leftrightarrow C_6H_{12}O_6 + 6O_2. \tag{8.25}$$

In daylight, during photosynthesis (8.25) runs from left to right consuming CO_2 and releasing O_2. At night, or during combustion, the direction is reversed and energy is released. Implicitly, the left hand side of (8.25) represents the oxidation of H_2O and the reduction of CO_2:

$$2H_2O \rightarrow 4H^+ + O_2 + 4e^-$$
$$CO_2 + 4H^+ + 4e^- \rightarrow (H_2CO) + H_2O. \tag{8.26}$$

These oxidation and reduction processes each require four photons (i.e. a total of eight), to convert each of six CO_2 molecules into glucose (8.25). The photosynthetically active region of the solar spectrum, $400\,nm < \lambda < 700\,nm$, peaks at a wavelength of $\sim 550\,nm$ which is equivalent to a photon energy of 3.6×10^{-19} J. The solar energy required at the peak wavelength to produce a carbohydrate molecule is therefore ($6 \times 8 \times 3.6 \times 10^{-19}$ or 1.73×10^{-17} J). Each carbohydrate molecule stores 4.66×10^{-18} J (8.25) yielding an efficiency in the order of 27%. This represents an ideal upper limit and in reality there are considerable losses from respiration and reflection that reduce the efficiency by a factor of ~ 5.

Most direct burning biomass power plants are based on the Rankine cycle (Sec. 6.3) and are designed to burn fast growing trees (e.g. willow) and/or sugar cane. Transportation costs dictate that the tree and cane plantations should be located close to the power stations which therefore tend to be small. Economies of scale are more difficult to realise and the use of high cost, high performance steels is difficult to justify. Consequently, station efficiencies are generally low and the electricity costs comparatively high.

Unlike coal, biomass fuels have high moisture content and a considerable fraction of the available energy is lost as water vapour. An alternative approach that has been used relatively successfully is to co-fire the biomass with a proportion of coal, taking advantage of both the cheaper cleaner biomass fuels and the higher operational temperatures attainable with coal to improve efficiency and reduce generating cost. In fact, most forestry and agricultural production devoted to fuel is converted to alcohol fuels for use in motor vehicles. It is the residues that are consumed in electrical power generation.

8.5.2 *Generation of electricity using biomass*

Virtually all agricultural and some solid municipal wastes can be used as feedstock for gas production, providing in the latter case a convenient and energy efficient means of waste disposal. The gasification of biomass fuels is a two step process as illustrated in Fig. 8.20. The biomass feedstock is first dried and then pyrolised at $\sim 400°C$ to produce a mixture of combustible (H_2, CH_4, CO + higher hydrocarbons) and incombustible gases (CO_2, N_2 and NO_x), solid carbon and ash. The carbon is then heated with steam, H_2 and O_2 at higher temperatures to produce more gas. The heat required for both steps is provided by the combustion of some of the carbon product. After purification, the resulting biogas may be used in a high efficiency combined-cycle gas fired turbine generating system to produce economic and carbon-neutral electrical power. If the process can be used to dispose of municipal wastes, then the double benefit of power production and reduced land-fill can, in principle, result in a net negative operational cost.

Another technological option is to convert the biomass using anaerobic digesters, where micro-organisms break down the material to produce CH_4. The technology is based on that already widely used by municipal authorities to transform sewage into innocuous products that may be safely disposed of or recycled. Although the CH_4 may be burned directly, it is more commonly used to generate H_2. This is discussed more fully in relation to transportation (Sec. 9.4).

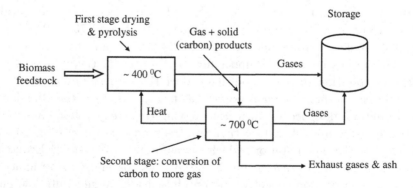

Fig. 8.20   Block diagram of gasification process for biomass fuels.

8.6 Hydroelectric Power

8.6.1 *Large scale hydroelectric power*

Hydroelectric power provides approximately a fifth of the global electrical energy supply; many orders of magnitude greater than other forms of renewable energy. Much of this is provided by large scale dam-based systems delivering 10's of MW of power well suited to grid-based networks — unlike most other forms of renewable energy. The electrical energy is of course derived from conversion of the gravitational potential energy of the water stored in the dam to kinetic energy of flow and using this, to rotate a turbine. In steady state, the flow leaving the dam must be replenished by the river(s) feeding into the dam. As a general rule, the river flow will vary depending on the season, being high in spring and lower in summer. On a longer term basis, it is important to know the 'flow duration curve' (FDC), which is the fraction of time that the flow is greater than a given (or design rated) flow. The FDC provides a useful indicator of the annual power that might be expected from a given hydroelectric facility, it does not, however, give any information as to how variable the supply from a given dam will be. Dams with larger storage capacity will be less susceptible to fluctuations in water flow than smaller hydro-schemes. Both however, are ultimately dependent on the rainfall over the catchment area and the character of the terrain.

Of their nature, large dam systems entail the flooding of considerable tracts of land. They also require suitably large rivers, capable of sustaining the outlet flows. About 90% of the power generated in Africa and South America derives from large facilities of this type. This has led to the formation of enormous artificial inland lakes, such as Lake Volta behind the Akosombo dam in the West African Republic of Ghana. Significant populations were displaced from their lands, causing considerable distress. The dam has undoubtedly provided much needed electrical power,[g] but there have been unintended and negative consequences. In the Ghanaian case, the waters of the lake have become infested with the debilitating bilharzia parasite infecting many of the lake-side communities with the disease. In 1994, work commenced on what will become the largest hydroelectric scheme in the world. The 'Three Gorges' project on the Yangtze river in China is intended to produce 18 GW of power from a 100 m high, 2 km wide dam. It is estimated that about a million people will be displaced by the 600 km lake expected to form behind the dam.

8.6.2 *Generation of hydroelectric power*

The output from a hydroelectric power station with an overall efficiency of η is given by:

$$P = \eta g \rho H Q \tag{8.27}$$

---

[g]As capacity often exceeds demand, Ghana is sometimes able to export electrical power to its neighbours.

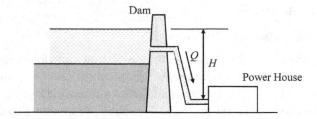

Fig. 8.21   Schematic diagram of a dam.

where H, ρ and Q are the head of water, the density and the volumetric flow $(\text{m}^3\text{s}^{-1})$ through the turbine respectively (Fig. 8.21). Expression (8.27) is a reasonable approximation for large systems where the head remains relatively constant and hence the potential energy of the water in the dam is unaffected by the proportion Q being drawn off to generate the power. In smaller scale systems, where replenishment does not match extraction, then H may vary and with it the power output. To some extent, this can be accommodated by controlling the flow rate, Q, to balance changes in H.

This is sometimes used to balance grid demand on very short time scales, as Q can be switched in or out very rapidly. In hydroelectric power stations used in this mode, the turbines are normally left freely running so minimising acceleration time. In some closed schemes, the turbines may be used either as power sources when demand is high, or when it is in low use spare capacity on the grid to pump the water back up into the dam reservoir. This is a convenient way of maintaining the steady operation of large coal-fired or nuclear facilities that cannot be so readily switched in or out to meet fluctuations in demand.

There are two basic types of turbine (Fig. 8.22): impulse turbines, where the water is directed through nozzles or jets onto bucket-shaped receptors located on the circumference of a wheel and reaction turbines akin to the propellers of a ship. Impulse turbines, sometimes referred to as Pelton turbines after their inventor (1880), convert the kinetic energy of the water into rotational motion and are therefore more suitable for use in high-speed, high-head, low-flow systems. Reaction turbines are entirely submerged in the flow and utilise the pressure difference across

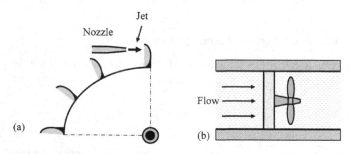

Fig. 8.22   Turbine types (a) impulse turbine; (b) reaction turbine.

appropriately designed runners and blades. They are best suited to low-head, high-flow conditions.

8.6.3 *Small scale hydroelectric and micro-hydro power generation*

While large scale hydroelectric plants can provide power on a scale comparable with a conventional fossil fuel power station, hydroelectric schemes may also operate at the small and even very small (micro-hydro) scales economically. Integration into a large national grid supply is not so convenient, but the environmental impact is considerably reduced; the flooding of large valleys is minimised and the use of 'run-of-the-river' type systems eliminate the need for a dam altogether. In these, some fraction of the river is siphoned off into a pipeline which delivers the water under pressure to a turbine situated downstream (Fig. 8.23). Since no dam is required, the ecological and environmental advantages are self-evident. However, micro-hydro schemes are only able to supply rather small amounts of power, more suited to use in isolated communities, e.g. farms, than for large industrial or urban conurbations.

The power supplied is critically dependent on frictional losses in the pipeline. The larger the diameter of the pipe, the lower the frictional loss, but the more expensive the installation becomes as the pipeline constitutes a major component of the construction costs. There are also issues to do with continuity of water supply, just as with major hydro-electric schemes, and given the small scale and absence of any effective storage capacity, micro-hydro power plants are particularly susceptible to variations of rainfall in the catchment area of the river. The friction loss can be represented as a reduction in the head at the turbine by replacing H the 'gross head' in (8.27) by $(H-H_F)$, where H_F is the 'friction head'. The frictional losses scale approximately as Q^2 and hence the net power delivered by a micro-hydro system would be:

$$P = \eta\rho g(H - H_F)Q = \eta\rho g(H - k_q Q^2)Q \qquad (8.28)$$

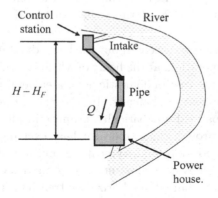

Fig. 8.23 'Run-of-the-river' micro-hydro power generation.

where k_q is a constant which includes the coefficient of friction, the length and diameter of the pipeline. It is evident from (8.28) that for a given type, diameter and length of pipe (i.e. k_q), there will be an optimum rate of flow through the system. If Q is too low, then the available power will be small, and if Q is too large, the frictional losses will be too great and the gross head will be unduly reduced. To determine the optimum value of Q for maximum power, we differentiate (8.28) w.r.t. Q:

$$\frac{dP}{dQ} = \eta \rho g(H - 3k_q Q^2) \tag{8.29}$$

and equate to zero:

$$H = 3k_q Q^2 = 3H_F \quad \text{i.e. } H_F = \frac{1}{3}H. \tag{8.30}$$

Condition (8.30) is a *design criteria*, not an *operational* one. It does not imply that Q should be controlled simply by partially shutting off or turning on valves, but that the piping should be chosen to match the anticipated mean flow. In reality, the design should be treated as an integrated whole, including the turbine, generator, expected fluctuations in the river flow and so on.

8.7 Ground Source Heat Pumps

Although the principles have been well known for some time, it is only comparatively recently, as operational problems have been addressed and reliability improved, that ground source heat pumps are a beginning to be widely used to provide hot water and environmental conditioning. The underlying basis is the thermodynamic heat pump described in Sec. 6.2.3. and illustrated in Fig. 6.2(b). Heat is pumped from the sub-surface of the earth, where the temperature is essentially constant (7°C–10°C in the UK) irrespective of the season (see Sec. 5.3.2), to provide domestic hot water. Work must be done to achieve this, but in a well designed ground source unit the heat delivered may be 2 to 3 times that consumed in the pump/compressor stages of the system.

There are several different schemes, but the block diagram, Fig. 8.24, illustrates the main principles of operation, at the heart of which is a conventional refrigerator.

Three closed (or in the case of domestic hot water, semi-closed) cycles exchange heat through the two heat exchangers. In the central refrigeration cycle, the refrigerant liquid is expanded adiabatically, evaporating it to produce a cold gas. The gas is passed up through the evaporator heat exchanger where it is warmed by the descending ground loop fluid, typically a mixture of water and antifreeze, which is cooled by an equivalent amount. The ground loop is usually a loop of polyethylene that has been buried in a horizontal trench or less commonly, (because of the expense) in a bore hole. As it flows through the ground loop, the chilled

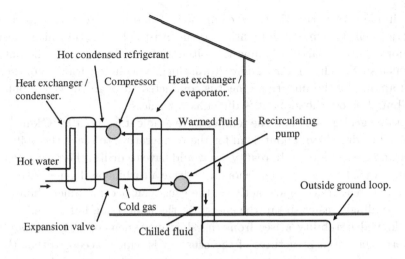

Fig. 8.24   Schematic diagram of a domestic ground source heating system.

water/antifreeze fluid extracts heat from the subsoil, is warmed and returned to the evaporator heat exchanger.

During its passage through the heat exchanger, the refrigerant temperature is increased from a few degrees below 0°C to a few degrees above. It is then fed to the compressor where it is liquefied, and its temperature increased to ~ 80°C, due to the work done on the gas and the release of heat of condensation. Thermal transfer in the condenser heat exchanger provides hot water (50°C–60°C) for the space heating and domestic hot water systems. In the latter case, the hot water is used to heat water in a conventional header tank configuration which must be properly insulated if the gains are to be fully realised.

Electrical energy is required to power the compressor and recirculating pump, but this can be significantly less than the heat delivered. However, although running costs are low, capital costs are high (digging trenches, installation of sophisticated heat exchangers etc.) and these systems are only cost effective over the long term. They are most effective as a source of space heating, where the hot water side is properly a closed system.

Reversing the direction of heat flow through the three cycles allows ground source systems to remove heat in summer, thus acting as air-conditioners.

8.8   Geothermal Energy

The sources of renewable energy discussed above are all ultimately derived from the sun. Geothermal energy is not and instead utilises the heat generated within the Earth's interior. Although this is a finite source like everything else, on human timescales, geothermal energy is for practical purposes limitless. It is carbon-free but there are other pollution issues to be resolved.

In simple terms, the Earth is composed of a solid core surrounded by two concentric shells known as the mantle and the crust. The mantle which constitutes the majority of the Earth's volume is molten and hot. In contrast, the outermost crust which sustains life, is only about 30 km thick, a negligible fraction of the whole. A small amount of the underlying heat is conducted through the crust, but as this is distributed across the globe it is difficult to harness.

In stable geological zones, the temperature gradient is about $30°C\,km^{-1}$ and we would need to drill down about 3 km for the temperature to reach the boiling point of water and extract steam. In reality, we would have to drill rather deeper to obtain steam that was, for example, hot enough to drive a turbine and generate electricity. However, in areas of geological activity, the crust can become much thinner and it is correspondingly easier to tap the energy in the underlying hot rocks.

Geological instability arises from the slow movement of large tectonic plates which form the major structures of the crust. It is widely accepted that the crust is composed of several such plates that move over the surface of the mantle. The processes are not well understood, but may be a consequence of slow cooling of the Earth's interior. Molten material from the mantle solidifies onto the surface of the core as a result, generating currents within the mantle on which the plates move. The boundaries where different plates meet, are areas of geological activity. Three major types of boundary are recognised: transform boundaries, divergent boundaries and convergent boundaries. Transform boundaries are those where the plates move past each other along a fault line and are characterised by major earthquakes. The famous San Andreas fault on the Californian coastline is a transform boundary between the Pacific and North American plates. Divergent boundaries occur when the plates are moving apart and are usually to be found mid-ocean (e.g. mid-Atlantic ridge), though not exclusively so, and are areas where new crustal material is being added from the interior of the Earth. Convergent boundaries are where the plates are moving towards each other, producing a thickening of the plates and the formation of mountain ranges such as the Alps and Himalayas. If one of the converging plates is oceanic, then the oceanic plate moves below the continental plate in a process termed as subduction. The subducted plate melts as it penetrates the mantle and less dense material rises to form volcanoes at the boundary.

Although there are a variety of geothermal sources, only hydrothermal sources are presently used commercially. Hydrothermal energy (Fig. 8.25), involves the extraction of steam from the site and using it to produce electrical power, or hot water for residential space heating. The high enthalpy sources, i.e. the steam sources, are associated with the edges of the main plates, in particular the 'Ring of Fire' Pacific rim (the Wairakei station in New Zealand and the major 'Geyser' installation in California) and the mid-Atlantic ridge on which Iceland is sited, are used in large scale power generation facilities. Hot water, low enthalpy sources are more common and are increasingly being utilised in district space heating schemes throughout Europe and Siberia.

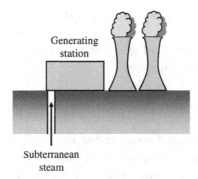

Fig. 8.25 Geothermal station based on the extraction of subterranean steam.

The power that can be extracted from a hydrothermal source is constrained by the amount of ground water and the rate at which it can be replenished, i.e. by rain or ground water diffusion. For example, the large Californian Geysers generating station was providing about 2 GW of power until in 1988, output started to reduce dramatically as ground water resources were used up and steps had to be taken to arrest the loss by injecting treated water from neighbouring towns.

The steam and hot water brought to the surface are heavily contaminated with mineral salts, particularly dissolved sulphur dioxide. The contaminants accelerate turbine blade corrosion and cannot be released to the environment. Their removal adds considerably to the operational costs of both geothermal power stations and district heating schemes.

8.9 Problems

1. A solar panel $0.75\,\text{m} \times 0.5\,\text{m}$ is fitted to a Mars probe. In orbit around the Earth, the module is found to deliver 75 W. Given that Mars is about 1.5 times as far from the sun as is the Earth, how much power would it deliver in orbit around Mars?

   (Earth solar constant $= 1370\,\text{Wm}^{-2}$)

2. An 'ideal' Si photovoltaic solar cell has a surface area of $1\,\text{cm}^2$ and a reverse saturation current density of $1.5 \times 10^{-13}\,\text{A cm}^{-2}$ at a temperature of 298 K. Under AM1.5 illumination the cell produces 14 mW of power and a short circuit current density of $30\,\text{mA cm}^{-2}$. Calculate the open circuit voltage, conversion efficiency and fill factor for the cell.

3. Derive the slope of the slope (dJ_0/dV_0) of the photovoltaic characteristic (8.7) for a 'real' cell. Hence show that if the shunt resistance (R_P) is much greater than the series resistance (R_S), i.e. $(R_P \gg R_S)$, then (dJ_0/dV_0) at open circuit is $\sim 1/R_S$.

4. A wind farm comprises three 50 m rotor diameter. The operators claim the site can deliver three million units (i.e. kW hr) of electrical power per annum.

Estimate the required minimum wind speed. (Assume that the density of air is $1.2 \, \mathrm{kg \, m^{-3}}$.)

5. The probability that the wind speed at a particular location is less than some value u_0 is given by Rayleigh statistics as:

$$P(u \leq u_0) = 1 - \exp\left\{-\frac{\pi}{4}\left(\frac{u_0}{\overline{u}}\right)^2\right\}$$

where $\overline{u}$ is the average wind speed at the site.

A wind turbine is rated to operate at wind speeds between $3.5 \, \mathrm{ms^{-1}}$ and $20 \, \mathrm{ms^{-1}}$. During one particular year of operation the wind speed was too low for the turbine to operate for a total of 804 hours. For the site in question, estimate (a) the mean wind speed and (b) the number of hours for which it was possible to operate the turbine.

6. A run-of-the-river micro-hydro power system takes $0.2 \, \mathrm{m^3 s^{-1}}$ of water from a small stream and delivers it through $300 \, \mathrm{m}$ of pipe to the turbine house which is situated $30 \, \mathrm{m}$ below the extraction point. If the turbine efficiency is 40%, and the optimum friction loss along the pipe is about $1.5 \, \mathrm{m}$ for every $30 \, \mathrm{m}$ of pipe, estimate the power that the system can generate.

References

1. D. M. Chapin, C. S. Fuller and G. L. Pearson, A new p-n junction photocell for converting solar radiation into electrical power, *J. Appl. Phys.* **25** (1954) 676.
2. See M. A. Green, K. Emery, Y. Hisikawa and W. Warta, Solar cell efficiency tables (Version 3), *Prog. Photovoltaics: Res. Appl.* **15** (2007) 425, for the most recent consolidated tables of cell efficiencies. These are updated on a regular and rigorously defined basis.
3. T. Carlsson and A. W. Brinkman, Identification of degradation mechanisms in field-tested CdTe modules, *Prog. Photovoltaics: Res. Appl.* **14** (2006) 213.
4. T. Burton, D. Sharpe, N. Jenkins and E. Bossanyi, *Wind Energy Handbook*, John Wiley & Sons (2001).

Chapter 9

TRANSPORTATION

9.1 Introduction

Mechanised transportation dates from the second half of the 18th Century, though it is really only with the advent of the railways in the early part of the 19th Century that it became widespread. Hitherto, transport on land had been based on the horse and at sea on sail: slow, limited in volume and uncomfortable even for the wealthy. The railways not only changed all of that, they contributed to the profound social changes of the Industrial Revolution. The rapid and economical transportation of raw materials to factories and finished articles to market, and the supply of agricultural produce from rural areas to feed the new expanding urban conurbations were all essential constituents of the new industrial society.

This was all powered by steam, and we shall consider briefly the principles of steam power and locomotion: reciprocating engines, flywheels and valves; all common place ideas now, but entirely novel at the end of the 18th Century. It is ironic that in the middle of the 19th Century, when steam powered travel was at its peak, the internal combustion engine which was to completely supersede steam was invented. Based on the burning of portable high energy density hydrocarbon fuels, the internal combustion engine made possible the development of the automobile and the era of private as opposed to public transport. Given its importance and ubiquitous nature, we shall discuss the internal combustion engine in somewhat greater detail; considering not only the mechanics and thermodynamics of its operation but some of the ecological issues as well.

Although the contemporary problems of emissions and fuel consumption associated with cars have serious implications for the global environment, the desire for people to retain the freedoms that personal transport brings requires the development of new and less damaging technologies. Foremost among these must be electrical vehicles, and consequently, we shall include a brief discussion of the physical and technical principles underlying the operation of electric motors. Unlike chemical energy, electrical energy is difficult to store and this presents a real challenge for the use of electric vehicles. Petrol and diesel fuelled vehicles can typically travel some hundreds of kilometres at high speed before refuelling. Battery driven cars are limited to a few tens of kilometres at rather more modest speeds before the need for a recharge. A promising (if expensive) route forward is the development of hydrogen powered engines. These are clean, producing only water vapour, although as a gas rather than a liquid — onboard storage of hydrogen is a serious impediment to their use. It is nevertheless probable that hydrogen fuel will play a significant role in the future, and many of the world's major motor manufacturers are currently working on hydrogen fuelled vehicles. We shall therefore also consider the use of hydrogen both as a replacement for petrol and diesel in conventional internal combustion engines and its use in fuel cell powered electric cars.

9.2 Steam Powered Transport

The invention of the steam engine is generally credited to Thomas Newcomen in 1698, and throughout the following century practical engines were developed by several engineers, among them James Watt. The earliest machines were designed for stationary applications, pumping water, steam foundries and the like. They were large contraptions employing great boilers and large pistons, quite unsuited for powered transport. Several attempts were made to develop a steam powered locomotive, but without success; the designs proved unreliable, boilers blew up, or they damaged the wooden rails. It was George Stephenson who a century later developed the first successful locomotive, the "Blucher", designed to haul coal wagons over short distances in the Tyne valley coalfields. Stephenson then proceeded to construct the first passenger rail link between Darlington and the nearby seaport of Stockton in 1825, a distance of about 32 km (20 miles) and in 1831, the Liverpool-Manchester railway, for which he designed and built his most famous and advanced locomotive the "Rocket". The transportation revolution of the 19th Century had begun. By the end of the 1830's, some 3000 km of railway had been laid and over succeeding decades, railway networks spread across Europe and America.

Motive power in steam engines is achieved through the reciprocating action of a piston moving in and out of a cylinder, controlled by the operation of valves, which regulate the ingress and exhaust of steam (Fig. 9.1(a)). The latter produced in the boiler under pressure is fed through the inlet valve, increasing the pressure in the chamber, and driving the piston out of the cylinder. The connecting rod transfers the displacement to the crank converting translational motion into rotational. At the end of the 'stroke', the exhaust valve is opened and the steam is let out of the cylinder. Simultaneously, a second valve admits steam to the other side of the piston forcing a return stroke.[a] A large flywheel connected to the crank shaft smoothes out variations in power delivery. The angular momentum imparted to the flywheel during power strokes acts as a store of mechanical energy to be released during quiescent parts of the cycle.

Work transfer in a reversible reciprocating single acting (for simplicity) engine is calculated by integrating around the cycle of the corresponding pV diagram (Fig. 9.1.(b)). This is not a proper thermodynamic cycle — it does not include,

---

[a]Engines of this type are sometimes referred to as *double action* engines.

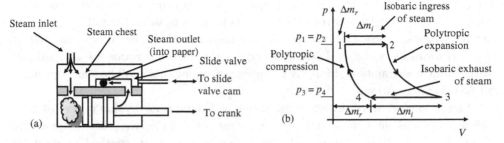

Fig. 9.1 Steam engine (a) schematic diagram of a double action engine; (b) $p-V$ diagram for single action engine.

for example, the steam heating part of the process — but will serve to illustrate the operation.

$1 \to 2$ Inlet valve is opened and a mass Δm_i of steam is admitted into the cylinder, where it mixes with the mass Δm_r remaining from the previous cycle, isobarically expanding the volume;

$2 \to 3$ Inlet valve is closed and the mass $(\Delta m_i + \Delta m_r)$ expands, driving the piston out of the cylinder;

$3 \to 4$ At the end of the expansion stroke, the exhaust valve is opened and a mass of steam Δm_i is ejected isobarically during the return troke;

$4 \to 1$ The exhaust valve is closed, the remaining mass of steam (Δm_r) is compressed and the cycle started again.

Unlike the turbines considered earlier (Sec. 6.2.4), steam throughput is slow and processes $(2 \to 3)$ and $(4 \to 1)$ are not isentropic, but *polytropic*; $pV^n = $ constant where n is the *index of expansion*. To calculate the net work done round the cycle, we compute the algebraic sum of all the work transfers:

$$\oint dW = \oint p dV = p_1(V_2 - V_1) + \frac{(p_3 V_3 - p_2 V_2)}{(1 - n)} + p_3(V_4 - V_3) + \frac{(p_1 V_1 - p_4 V_4)}{(1 - n)}.$$

(9.1)

Collecting terms and remembering that $p_1 = p_2$ and $p_3 = p_4$ we get:

$$\oint dW = \frac{n}{(1 - n)} \{p_1(V_1 - V_2) + p_3(V_3 - V_4)\}.$$

(9.2)

We may express the volumes in terms of the *specific volumes* ($v = V/m$, i.e. the inverse of the density), thus

$$V_1 = v_1 \Delta m_r; \quad V_2 = v_1(\Delta m_r + \Delta m_i); \quad V_3 = v_3(\Delta m_r + \Delta m_i); \quad V_4 = v_3 \Delta m_r.$$

(9.3)

Substituting (9.3) in (9.2) gives the work done per cycle:

$$\oint dW = \Delta m_i \frac{n}{1 - n}(p_3 v_3 - p_1 . v_1).$$

(9.4)

Equation (9.4) implies that we need $p_3v_3 > p_1v_1$ in order to maximise the work done per cycle; the piston stroke should be as large as possible (large V_3/V_1) within the limitations imposed by the physical dimensions of the locomotive as well as considerations of mechanical strength and weight. Pressure ratios are also subject to the same constraints and for any given size of engine there will be some optimum set of design conditions.

Steam engines were invented and developed some considerable period of time before there was any conceptual understanding of the underlying thermodynamics. Arguably, it was the success of the steam engine that sparked interest in thermodynamics and led to such notions as the equivalence of heat and mechanical energy and the fundamental laws of thermodynamics (Sec. 6.2). They were also invented at a time when coal was the only available fuel. Coal burning locomotives continue to be used in some parts of the world, even though they are highly polluting, emitting large volumes of particulates (soot). Since the heat source is external to the boiler, other cleaner fuels could be used, but these are better utilised in the internal combustion energy, as we shall discuss next.

9.3 The Internal Combustion Engine

9.3.1 *The Otto cycle petrol engine*

The Internal Combustion Engine (ICE) has antecedents in steam in that both utilise reciprocating pistons in cylinders. However, the similarities end there. The source of heat used to produce steam in a steam engine is, in principle, unimportant. This is not so in the IC engine — as anyone who has inadvertently filled their fuel tank with the wrong variety of fuel will testify. An ICE designed to run on diesel will not do so on petrol. Although some gas fuelled cars were constructed in the middle of the 19th Century, it was not until Otto and Diesel developed ICEs based on petrol and diesel respectively, that cars became a practical proposition. This together with the increasing availability of refined oil products meant that by the end of the first decade of the 20th Century, cars had become reasonably common throughout Europe and America.

The basic structure of the Otto (petrol) engine is illustrated in Fig. 9.2(a). In both petrol and diesel engines, the full cycle is carried out over two or more commonly four strokes of the piston (Fig. 9.2(b)).

$1 \rightarrow 2$ *Induction Stroke*: Fuel/air mixture is drawn in through the inlet valve as the piston moves out. The inlet valve is closed at the end of the stroke.

$2 \rightarrow 3$ *Compression Stroke*: The fuel/air mixture is compressed as the piston moves back in. Just at the end of the stroke, the mixture is ignited such that combustion takes place under approximately constant volume conditions.

$3 \rightarrow 4$ *Expansion or Work Stroke*: The sudden impulse of heat drives the piston out, producing useful work in the process.

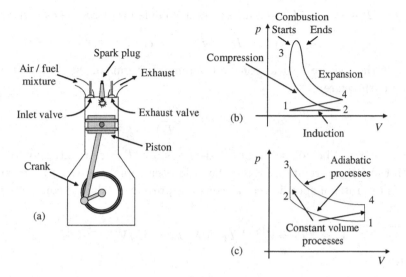

Fig. 9.2   Petrol ICE (a) schematic diagram; (b) four stroke cycle; (c) p–V diagram.

$4 \to 1$ *Exhaust Stroke*: The exhaust valve is opened allowing exhaust products to be displaced as the piston moves back up the cylinder, to restart the cycle.

The four stroke sequence is well suited to the four cylinder format generally used in ordinary domestic vehicles. The four cylinders are timed so that one is always producing work, driving piston movements in the other three cylinders and smoothing out the supply of power.

The Otto cycle is not strictly a closed cycle since the exhaust is not recirculated. Nevertheless, the approximation is a good one (and certainly better than for steam) since the air/fuel ratio is large (~ 15) and of course air is predominantly N_2 which does not play a significant part in the combustion process. Thermodynamically, the Otto cycle may be represented by the simplified p–V diagram in Fig. 9.2(c).

$1 \to 2$ Air-fuel mixture is compressed from V_1 to V_2;

$2 \to 3$ The fuel is ignited by an electric spark. Combustion effectively takes place under constant volume conditions providing an impulse of heat that results in a sudden and large increase in the cylinder pressure, from p_2 to p_3;

$3 \to 4$ The increased cylinder pressure drives the piston out of the cylinder isentropically, providing useful work in the process;

$4 \to 1$ Heat Q_{41} is rejected at constant volume to complete the cycle.

Defining the efficiency of the (reversible) process as (6.14):

$$\eta = \frac{W}{Q_i} = \frac{Q_i - Q_o}{Q_i} = 1 - \frac{Q_o}{Q_i} \tag{9.5}$$

where W, Q_i and Q_o as the useful work done, the heat supplied (i.e. by ignition of the fuel ($2 \to 3$)) and the heat rejected to the environment, ($4 \to 1$) respectively.

Since both $(2 \to 3)$ and $(4 \to 1)$ are constant volume processes we can write:

$$Q_i = c_v(T_3 - T_2) \quad \& \quad Q_o = c_v(T_4 - T_1) \tag{9.6}$$

where c_v is the specific heat at constant volume. In terms of the temperatures the efficiency is therefore:

$$\eta = 1 - \frac{(T_4 - T_1)}{(T_3 - T_2)}. \tag{9.7}$$

For a constant number of moles of an ideal gas, $pV/T = $ constant and combining this with the Poisson relation for an adiabatic (isentropic) process, $pV^\gamma = $ constant where γ is the ratio of specific heat at constant pressure to that at constant volume, we obtain:

$$V_3^{(\gamma-1)} T_3 = V_4^{(\gamma-1)} T_4 \quad \text{or} \quad T_4 = (V_3/V_4)^{(\gamma-1)} T_3. \tag{9.8}$$

Similarly;

$$T_1 = (V_2/V_1)^{(\gamma-1)} T_2. \tag{9.9}$$

However, $V_1 = V_4 = V_{\max}$ and $V_2 = V_3 = V_{\min}$ (fig. 9.2(c)). We define the ratio $V_{\max}/V_{\min}$ as the *compression ration*, r_p. Substituting for the temperatures in (9.7) and expressing the result in terms of the compression ratio:

$$\eta = \frac{T_3(1 - (1/r_p^{(\gamma-1)})) - T_2(1 - (1/r_p^{(\gamma-1)}))}{(T_3 - T_2)} = 1 - \frac{1}{r_p^{(\gamma-1)}}. \tag{9.10}$$

Expression (9.10) clearly implies that the greater the compression ratio, the higher the efficiency, except that at compression ratios of ~ 10, air-petrol mixtures explode spontaneously, causing 'pinking' in improperly timed engines. Spontaneous combustion can be suppressed to some extent by the use of additives such as tetra-ethyl lead. However, the environmental and health implications of this compound subjected it to being banned in many countries. Other supposedly less harmful additives have been developed (so-called lead-free petrol) which combined with better engine design, has meant that efficiencies have continued to improve.

The efficiency defined in (9.5) is for a reversible closed system. Real engine cycles are neither reversible nor, as pointed out above, an ideally closed one. In operation, some work always has to be expended in overcoming the viscosity of the lubricants, as well as running ancillary components, such as the valve gear, water and oil pumps, alternator, etc. Thus the work actually available to drive the vehicle is somewhat less than that implied in (9.5).

9.3.2 *The Diesel cycle engine*

The Otto cycle engine efficiency can be improved by increasing the compression ratio, but only to the point where the fuel-air mixture explodes spontaneously. In the

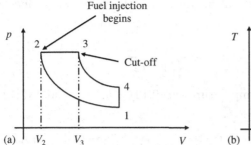
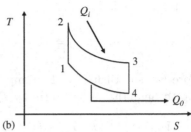

Fig. 9.3   Diesel engine cycles (a) $p-V$ diagram; (b) $T-S$ diagram.

Diesel cycle, this is resolved by compressing the air first and only then injecting the fuel when the air in the cylinder is at maximum compression, initiating combustion without the aegis of a spark plug. The basic diesel engine, the structure of which is essentially the same (apart from the spark plug) as the petrol engine (Fig. 9.2.(a)), is also based on a sequence of four operations:

(i)   Induction;
(ii)   Compression;
(iii)   Injection and combustion at maximum compression, producing expansion (i.e. work output);
(iv)   Exhaust.

The corresponding $p-V$ and $T-S$ diagrams are illustrated in Fig. 9.3. Air is compressed adiabatically from V_1 to V_2; V_1/V_2 is the compression ratio r_p as in the Otto cycle. Fuel injection starts at 2 and since the pressure is far in excess of the combustion threshold, ignition starts immediately. Fuel injection and therefore combustion take place over a finite period of time and continue into the expansion stroke, and hence occur under isobaric (represented by the horizontal line on the $p-V$ diagram) rather than constant volume conditions. When the piston reaches position 3, fuel injection is cut off. We define a *cut-off ratio*:

$$r_{cf} = \frac{V_3}{V_2} \tag{9.11}$$

which we can express using the ideal gas law and remembering that $p_2 = p_3$:

$$\frac{V_2}{T_2} = \frac{V_3}{T_3} \quad \text{or} \quad r_{cf} = \frac{V_3}{V_2} = \frac{T_3}{T_2}. \tag{9.12}$$

After the cut-off, the piston continues to move out adiabatically (no more heat is being added) to 4 and between 4 and 1, heat Q_o is rejected to the environment at constant volume (9.6). The heat absorbed during combustion (2→3) is:

$$Q_i = c_p(T_3 - T_2) \tag{9.13}$$

where c_p is the specific heat at constant pressure. The thermodynamic efficiency is defined in the usual way:

$$\eta = 1 - \frac{Q_o}{Q_i} = 1 - \frac{T_4 - T_1}{\gamma(T_3 - T_2)}. \tag{9.14}$$

As before, we substitute for the temperatures in (9.14) using (9.8) and (9.12) and dividing through the second term by T_2:

$$\eta = 1 - \frac{(V_3/V_4)^{(\gamma-1)}(T_3/T_2) - (V_2/V_1)^{(\gamma-1)}}{\gamma((T_3/T_2) - 1)} = 1 - \frac{(V_3/V_4)^{(\gamma-1)}r_{cf} - (1/r_p)^{(\gamma-1)}}{\gamma(r_{cf} - 1)}. \tag{9.15}$$

Note we can express the ratio (V_3/V_4) as:

$$\frac{V_3}{V_4} = \left(\frac{V_3}{V_2}\right)\left(\frac{V_2}{V_4}\right) = \frac{r_{cf}}{r_p}. \tag{9.16}$$

Substituting this into (9.15), we obtain for the efficiency

$$\eta = 1 - \frac{1}{r_p^{(\gamma-1)}} \times \left\{\frac{1}{\gamma}\frac{(r_{cf}^{\gamma} - 1)}{(r_{cf} - 1)}\right\}. \tag{9.17}$$

The expression for the efficiency of a diesel engine differs from that for a petrol engine (9.10) by the bracketed factor in the second term. A diesel engine will be more efficient than an equivalent petrol engine because the compression ratio can be increased, but this advantage is reduced by the bracketed term. Since $r_{cf} > 1$, the bracketed term is also greater than unity and increases slowly with r_{cf}, progressively reducing the efficiency. We therefore wish to keep the bracketed term small, but doing so reduces the work per cycle (cf problem with Carnot engine), and in practice both the compression and the cut-off ratios have to be adjusted to optimise the design of a particular engine.

9.3.3 *Emissions and catalytic converters*

The burning of fossil fuels, whether in ICEs or in power stations, always results in the production of CO_2, which although of major importance as a greenhouse gas, has few other health hazards. In practice CO_2 is seldom the only emission, and others are often of more immediate concern to human and plant health. Much of this hazardous pollution comes from road transport where the conditions under which the engines must operate vary significantly and where available technologies are difficult to incorporate within reasonable cost margins. Similar emissions will also be produced by large scale power stations, but there cost margins are less of a constraint and regulation easier to enforce.

The burning of hydrocarbons (C_xH_y) in air will produce a mixture of gaseous emissions: CO_2, CO, NO_x and volatile organic compounds (VOC) together with any other emissions associated with impurities in the fuel, such as SO_2 and particulates. An equilibrium will always exist between CO_2, CO and O_2 according to the chemical equation:

$$CO_2 \leftrightarrow CO + CO + \frac{1}{2}O_2. \tag{9.18}$$

The reaction from left to right is endothermic, i.e. heat must be extracted from the environment for it to occur, i.e. for CO_2 to be converted into CO and O_2. Carbon monoxide is a poisonous gas that binds to oxygen receptors in blood haemoglobin, preventing it from transporting oxygen making the victim suffocate. Reaction (9.18) is a not infrequent cause of fatal accidents in garages and from faulty heating appliances in the home.

Nitrogen oxides, principally a mixture of NO and NO_2 and collectively written, NO_x will be produced in high temperature reactions between N_2 and O_2. Since ICEs burn fuels in air, then NO_x will always be produced. Neither NO nor NO_2 pose a direct hazard at ordinary concentrations, but are major contributors to the production of photochemical smog. The reaction between VOCs and NO_x in the presence of sunlight produces ozone which while desirable in the stratosphere is not so at ground level (Sec. 4.5).

Particulates or aerosols are small particles ($< 100\,\mu m$ in size) of matter that are suspended in the air. The distribution of particle sizes tends to be bimodal: a fraction distributed about $\sim 20\,\mu m$ and a second smaller submicron sized fraction. It is the latter that constitutes the greater health hazard, since it penetrates more deeply into the lungs. Regulations usually refer to the mass per cubic metre ($\mu g\,m^{-3}$) of particles less than 10 μm in size, referred to as PM_{10}. Although similar particle distributions seem to be part of the natural environment, and are observed in even the remotest rural locations, the number density (i.e. number of particles per unit volume) is very much greater in urban areas, and the chemical makeup is different. Aerosols in rural areas are composed mostly of silicates, those in city environments have considerably larger concentrations of carbon and metal particles, of which much is produced by traffic. Major installations such as power stations have cyclonic and electrostatic devices for the removal of particulate emissions, but these are impractical for use in vehicles.

If, ultimately, the production of CO_2 from the burning of hydrocarbon fuels cannot be avoided, some of the other harmful emissions can be removed or reduced by improved technology. The endothermic chemical equilibrium (9.18) between CO and CO_2 can be driven entirely to the CO_2 side and the analogous NO_x endothermic reaction reversed to produce N_2 and O_2 at the appropriate equilibrium temperature. Under ordinary exhaust conditions there is insufficient time for equilibrium to become established, but introducing a suitable catalyst into the tailpipe can speed up the process. In cars, this has led to the development of the *three-way catalyst* which catalyses the oxidation of CO to CO_2, uncombusted C_xH_y to CO_2 and H_2O, and reduces NO_x to N_2 and O_2. The catalysts typically employ large surface areas of precious metals, e.g. Pt, Pa, and Rh which are not consumed in the process, though they may become contaminated.

9.3.4 *Alternative hydrocarbon fuels*

A variety of alternative fuels for cars have and are being developed, with a view to reducing or eliminating pollution, including biomass and gas fuels. Probably the most widely used is Liquid Petroleum Gas (LPG), which burns cleanly and emits no particulate emissions from the combustion process apart from particles due to engine wear and lubricating oils. The flame temperature is lower and thus NO_x emissions are correspondingly reduced. A distribution infrastructure has already been established in Western Europe and parts of the USA.

Ethanol fuels derived from biomass are particularly interesting. An alcohol based on ethane in which one of the hydrogen atoms has been replaced by an OH bond, ethanol burns cleanly and since it is produced from biomass, is carbon neutral. The CO_2 released simply being that taken up by the plant during its life cycle. Ethanol is sometimes used as an admixture with petrol (gasoline) to produce *gasohol* which is better suited to standard engine designs. The other alcohol fuel methanol does not burn as cleanly because there is some production of formaldehyde (H_2CO) which is toxic. Methanol can be produced from methane released by the decay of biomass, in which case it is carbon neutral as well.

9.4 Hydrogen powered vehicles

9.4.1 *The hydrogen economy*

The use of hydrogen as a clean fuel has long been recognised[1] — as have many of the technical problems associated with its use. Hydrogen is not a *primary* source of energy; unlike oil and gas, the element is not freely available in nature but has to be generated from some other primary energy source, i.e. hydrogen is a *secondary* source. The extent to which it is a clean technology depends on the 'cleanliness' of that other source. Obviously, if the hydrogen is produced from say, the electrolysis of water using electricity generated from a fossil-fuelled power station, then not only is it polluting, but inefficient as well! It would be better to use the electricity generated by the station directly. If on the other hand, hydrogen is produced in an integrated way from renewable sources, then it provides a very attractive way of smoothing out the variability of energy supply and matching it to demand. Its use in vehicles poses additional problems of storage and lack of a distribution infrastructure.

The integrated use of hydrogen across the board in both large utilities and small yet to be realised personal cars is generally understood to be the 'Hydrogen Economy'. Many perceive it as the successor to the present day 'Oil Economy'. As a gaseous chemical fuel, it is relatively easy to use as a hydrocarbon replacement in internal combustion engines, gas turbines and externally heated systems. It also offers the unique option of direct conversion to electrical energy in a fuel cell with a substantially improved efficiency. What is more, as a low mass, low viscosity gas, hydrogen is easily piped over long distances without expending disproportionate amounts of power.

There are, of course, downsides to its use. We have already alluded to the problems of onboard storage in vehicular use. There are several possible technologies currently under research and development, but none has so far become sufficiently superior to its competitors to warrant adoption as the industry standard. This in itself is a major drawback, since the immense capital investments required in setting up a distribution network will not be made until there is general agreement as to the preferred mode of storage.

The use of hydrogen in an ICE requires only changes to air/fuel mixtures and timing etc. The principal product is water vapour, though in this type of application some NO_x will also be produced, because combustion takes place in air. No nitrogen oxides are produced in the fuel cell described next.

9.4.2 *The hydrogen fuel cell*

Fuel cells are not new. The first report of a fuel cell was in a paper by Sir William Grove in 1839 where he described an experiment in which the chemical reaction of hydrogen and oxygen could be made to produce what he referred to as a 'gaseous voltaic battery' capable of inducing a painful shock in an individual. In 1842, he demonstrated to the Royal Society the use of such a cell (Fig. 9.4) to electrolyse water.

Although Grove originally foresaw the potential of his cell, it was not pursued as a viable source of energy partly because the feedstock gas (H_2) was both rare and expensive at the time, and partly due to corrosion effects resulting in short cell life. It was not for another half a century, that there was any further progress, when Mond

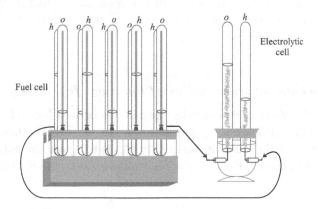

Fig. 9.4   Sir William Grove's 'Gas Voltaic Battery'.[2]

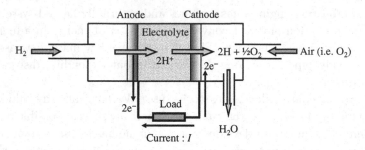

Fig. 9.5   Operation of a double chamber PEM cell.

and Langer developed a 50% efficient, 1.5 V cell. It was they who coined the term 'fuel cell'. The first practical fuel cell is generally credited to Bacon (a descendent of the 17th century scientist) who developed a 5 kW alkaline cell. However, it was the demands of space flight, and the need to power onboard instrumentation systems that really stimulated fuel cell development. An added bonus for the early manned flights (Gemini and Apollo series) was that the pure water produced by the fuel cells could be used as drinking water by the crew.

The operation of the fuel cell depends on the action of a permeable membrane that allows the passage of ionic charges of one sign but blocks the transport of electrons. Where the ions are protons (i.e. H^+), the membrane is referred to as a proton exchange membrane (PEM). Figure 9.5 illustrates the operation of a PEM cell. Hydrogen is supplied from the left and is ionised at the anode:

$$H_2 \leftrightarrow 2H^+ + e^-. \tag{9.19}$$

There are, in consequence, concentration gradients of protons (H^+) and electrons (e^-) across the membrane, but only the protons can diffuse through the membrane, electrons encounter a barrier and the resulting separation of charge sets up an e.m.f. The electrons must flow round an external circuit producing useful work 'enroute' to discharge the H^+. Hydrogen reacts with oxygen (in the air) in the right hand chamber to produce water vapour as the exhaust and maintain the concentration gradient across the membrane.

9.4.3 *The electrical characteristics of hydrogen fuel cells*

The oxidation of hydrogen, whether by combustion or in a fuel cell, is exothermic and releases stored chemical energy: the enthalpy change (ΔH) is therefore negative. We can use (6.9) and the Clausius inequality (6.11) to relate the enthalpy released by the reaction and the heat rejected δQ by the process to the electrical work done ΔW_E:

$$\Delta W_E \leq -(\Delta H - T\Delta S). \tag{9.20}$$

The bracketed term on the right of (9.20) is the change in Gibb's Free Energy (ΔG) and may be thought of as the 'entropy-free' part of the chemical energy that can be converted directly (in this case) into electrical energy. The efficiency is just:

$$\eta \le \frac{-(\Delta H - T\Delta S)}{-\Delta H} = \frac{\Delta G}{\Delta H}. \tag{9.21}$$

Values of ΔG and ΔH can be obtained from tables at standard conditions and for a hydrogen fuel cell, the maximum efficiency is $\sim 83\%$.

In order to determine the voltage produced by a cell, we need to determine the Gibb's free energy per ion ($\Phi\,eV$) and relate that to the fuel consumed. For a consumption of n moles, the Gibb's free energy is

$$\Delta G = (nN_A)e\delta\Phi \quad \text{or} \quad \Phi = \frac{\Delta G}{nN_Ae\delta} \tag{9.22}$$

where N_A is Avogadro's number and δ is the number of electrons exchanged in the reaction; the product (nN_A) is therefore the number of ions consumed. The power (P) delivered by a cell in operation will be:

$$P = \frac{d\Delta G}{dt} = \Phi e\delta N_A \frac{dn}{dt} = I \times V. \tag{9.23}$$

The current (I) flowing through the circuit must equal the number of moles of ions flowing through the membrane/sec. The voltage is therefore under the specified operating conditions:

$$V = \frac{P}{I} = \frac{\Phi e\delta N_A(dn/dt)}{eN_A(dn/dt)} = \delta\Phi. \tag{9.24}$$

Under open circuit conditions, $V = \delta\Phi$, the Gibbs free energy of formation of (liquid) water is $237\,\text{kJ mole}^{-1}$, $\delta = 2$, and from (9.22) we find that the theoretical voltage is $V = 1.23\,\text{V}$.

It has been assumed in (9.23) and (9.24) that the cell can supply the required current, i.e. that the ion current dn/dt can be supplied. In practice, as the current increases, the cell voltage will decrease. Real cells suffer a variety of internal losses arising from inherent factors such as the energy consumed at the electrodes to initiate the reactions (known as activation loss), Ohmic loss due to the resistance of the membrane to the passage of the ion current as well as more external effects associated with the transport of gases to and from the electrodes. For example, if exhaust water is not removed rapidly enough the supply of fresh air (O_2) will be impaired. The open circuit voltage in a real fuel cell is, as a result, typically reduced to about 1 V and under increasing load, the output voltage is progressively decreased. To a first approximation, fuel cells may be modelled rather like a battery, as an ideal voltage source in series with an internal resistance.

9.4.4 *Types of fuel cell*

Sir William Grove's fuel cell used graphite electrodes and H_2SO_4 as the electrolyte, which at that time was impracticable. Modern fuel cells encompass a wide variety of electrolytes and catalytic electrodes, but may broadly be classified in terms of the ion involved. In the discussion so far we have tacitly assumed that it is the hydrogen ion that forms the basis for the cell. This is partly for historical reasons, but also because whichever ion is used, the fuel is usually H_2. In practice, any of the ions: H^+, OH^- and O^{2-} may be used.

There are practical constraints on the electrolyte or membrane used in a given cell. Ideally, it should be:

  (i)   Mechanically strong;
 (ii)   Chemically resistant to impurities;
(iii)   Affordable;
 (iv)   Discriminate effectively against electron flow;
  (v)   Impede the passage of fuel.

Conditions (i) → (iii) are self evident, and (iv) should be plain from the discussion. The need (v) to prevent the passage of fuel through the membrane is less obvious. Hydrogen that crosses through the membrane without releasing its electron to the external circuit is wasted and contributes to internal losses in the cell (it is rather like the leakage current in solar cells, see Sec. 8.2.2).

The first systematic research into suitable proton exchange membranes was undertaken by NASA scientists in the 1960's, seeking to produce a viable fuel cell for use in space flights. They produced a sulphonated polymer by complexing with sulphonic acid. The sulphonated sites proved to be good H^+ ion conductors, but the polymer was mechanically weak. Subsequent work by DuPont, where the sulphonated polymer was interspersed with a fluorinated polymer, produced a much stronger block copolymer they called Nafion. As a film forming plastic, Nafion is easily processed, but is sensitive to impurities, particularly CH_4 (a potential source of H_2) and at \$600 per m^2, expensive. Like most polymers, it is a low temperature material and operational temperatures must be kept below about 80°C if the membrane is not to be degraded.

Alkaline fuel cells (the ones used by Bacon in his research) are based on transport of the OH^- ion and were used on the Apollo and Space Shuttle programmes. These cells used potassium hydroxide (KOH) as the electrolyte and porous (i.e. large surface area) Ni electrodes. The corresponding electrochemical reactions are:

$$\text{Cathode} \quad \frac{1}{2}O_2 + H_2O + 2e^- \leftrightarrow 2OH^- \tag{9.25}$$

$$\text{Anode} \quad H_2 + 2OH^- \leftrightarrow 2H_2O + 2e^- \tag{9.26}$$

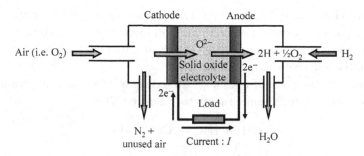

Fig. 9.6   Operation of a solid oxide fuel cell (SOFC).

These are highly efficient cells, but unfortunately, KOH is very reactive with CO_2 and thus these cells cannot easily be used in terrestrial conditions with air as the source of oxygen.

Oxygen ion transport forms the basis of the Solid Oxide Fuel Cell (SOFC). The electrolyte in a SOFC is a solid oxide ceramic. It is therefore strong, chemically and thermally stable and as such, may have a longer useable lifetime than other types. The operation of the SOFC is illustrated in Fig. 9.6; oxygen reacts at the cathode with electrons which have flowed round the external circuit to form the O^{2-} ions, which then diffuse through the ceramic electrolyte to the anode where they recombine with H_2 to form water by releasing two electrons.

$$\text{Cathode} \quad \frac{1}{2}O_2 + 2e^- \leftrightarrow O^{2-} \tag{9.27}$$

$$\text{Anode} \quad H_2 + O^{2-} \leftrightarrow H_2O + 2e^- \tag{9.28}$$

Though compact, SOFCs are not ideally suited for use in vehicles, because they operate at high temperature, ($\sim 750°C$) and repeated thermal cycling leads to fracture. They are much more suitable for use in large power station facilities, involving continuous operation and where the heat production can be used more effectively in cogeneration using CH_4 fuels. The requirement to discriminate against electrons is difficult to achieve in the high temperature, highly reducing environment in which SOFCs have to operate and this restricts the choice of ceramic considerably. Oxygen also has a rather large ionic radius (~ 1.4 Å) and the ceramic must therefore have a relatively open structure with available sites for mobile O^{2-} species to use. These conditions are well met by the so-called fluorite structured oxides (the term refers to the crystal structure and not the composition) of which Gd doped CeO_2 is the archetype. Other widely used ceramics are those based on the sintered zirconia-yttria (ZrO_2–Y_2O_3) materials systems.

9.4.5 *Production of hydrogen*

As pointed out earlier, hydrogen is not a naturally occurring fuel, it must be manufactured in some way. This adds to the cost and complexity of hydrogen use. The principal technologies currently in widespread use are steam reforming of methane, partial oxidation of methane and electrolysis of water. The first two processes utilise a hydrocarbon fuel as the initial source and are not carbon neutral unless the CH_4 is derived from biomass.

In the steam reforming of methane[b] a mixture of methane and steam is passed through a catalyst at high temperature to produce CO and H_2:

$$CH_4 + H_2O \leftrightarrow CO + 3H_2. \tag{9.29}$$

The reaction is endothermic and typically the required heat is supplied by burning a fraction of the methane or by burning the CO in a water-gas reaction which in addition to being exothermic, has the added advantage of increasing the H_2 concentration:

$$CO + H_2O \leftrightarrow CO_2 + H_2. \tag{9.30}$$

Residual quantities of CO remain and must be removed if the H_2 is to be used in PEM or alkaline cells. Steam reformed CH_4 is particularly well suited for use in SOFCs where if the cells are used on-site, thermal energy released in (9.30) can be used to heat the ceramic electrolyte.

Partial oxidation of CH_4 is an alternative method of breaking down the gas into H_2 and CO constituents. The reaction

$$CH_4 + \frac{1}{2}O_2 \leftrightarrow CO + 2H_2 \tag{9.31}$$

is exothermic and so will produce its own heat. The concentration of H_2 in the resulting gas may be increased using the same water-gas reaction (9.30).

Although both steam reforming and partial oxidation of methane create CO and CO_2, if the methane is produced from the gasification of biomass and municipal wastes, the technologies can be carbon neutral. Even when the CH_4 is derived from fossil fuel sources, either directly as natural gas or by the gasification of coal, the technology does significantly reduce emissions.

The electrolysis of water is simply the reverse fuel cell reaction. The load is replaced by an appropriate power source and the reaction driven in the opposite direction (Fig. 9.7). Hydrogen produced by electrolysis is highly pure and suitable for use in PEM and alkaline fuel cell systems susceptible to carbon based impurities. Production efficiency is high, but requires a source of electrical power and if this is provided by a renewable source, then it becomes a genuine source of carbon free energy, available whenever needed (Fig. 9.8).

One proposed scheme, the Wasserstoff-Energie-Island-Transfer (Hydrogen-Energy-Iceland-Transfer) aims to ship hydrogen produced in Iceland using renewable geothermal and hydroelectric sources of energy, which Iceland has in

---

[b]The steam reforming of methane is commonly used in the production of H_2 for use in other industrial processes such as the production of ammonia.

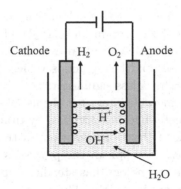

Fig. 9.7   Electrolysis of water.

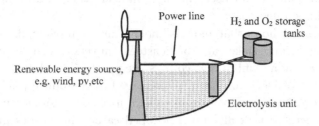

Fig. 9.8   Generation of hydrogen by electrolysis using renewable energy.

abundance to Northern Germany. While still in the early stages of development, such ideas may prove to be the way of the future.

9.4.6 *On-board hydrogen systems in vehicles*

Broadly speaking, hydrogen may be stored as a cryogenic liquid, a pressurised gas, absorbed on a metal hydride or adsorbed in carbon nanotubes. All of the modes of portable storage have serious technical difficulties and some advantages. The benchmark against which the efficient storage of hydrogen (or any other fuel) has to be compared remains as petrol. In a typical modern family car, the fuel tank holds about 40 kg of petrol giving a range of ~ 550 km. Any onboard fuel system has to be able to match this, without undue sacrifice of storage space.

Cryogenic systems clearly require equipment to liquefy the gas, expensive high quality insulation to minimise boil-off, and fail-safe valve systems to vent gas boiled-off. However, cryogenics is a well established technology and it is probably the case that none of these pose insurmountable difficulties. Although energy must be expended in liquefaction of the gas, supply logistics are probably simpler and with automated pump systems delivering a tank full of fuel in about the same time as for conventional hydrocarbon fuels, cryogenic hydrogen has much to commend it. It has been the hydrogen fuel of choice for BMW during its development of a hydrogen car.[c]

---

[c]In 2000, BMW deployed fifteen cars using ICEs and liquid H_2 as the fuel.

The storage of hydrogen as a pressurised gas has the major disadvantage that a pressure vessel is required to contain the gas. While the weight penalty has severe implications for the use of compressed hydrogen in private cars, it remains an attractive option for buses and heavy goods vehicles, where the weight of the compressed hydrogen cylinders is less significant.[d]

Initially, there was considerable interest in the use of metal hydrides as a means of storing hydrogen. The metal hydrides, typified by titanium iron hydride, absorb hydrogen, releasing some heat (heat of formation) in the process. When sufficient heat (heat of decomposition) is supplied, the process is reversed and hydrogen is released. Although hydride technology has steadily improved, storage efficiencies are low, typically a few percent by weight. They also suffer from poor kinetics and short lifetimes. However, metal hydride storage offers one singular advantage, that it is intrinsically safe. In an accident, it is unlikely that significant quantities of hydrogen would be released.

There was some hope that nanotechnology might provide the solution when in 1998, Rodriguez and Baker at Northeastern University claimed that they had developed a graphite nanofibre that could store up to 65% of hydrogen by weight.[3] Unfortunately, no other group have yet managed to reproduce (and hence verify) their work, provoking some scepticism about nanotube storage systems. Recent research has suggested that alkali-metal doped carbon nanotubes may be able to store up to $\sim 15\%$ by weight.

In addition to the problems of onboard hydrogen storage and the lack of a distribution infrastructure, fuel cell powered vehicles are very expensive. The electrical motor for a fuel cell vehicle costs approximately the same as a conventional internal combustion engine. In addition, the fuel cells themselves presently cost two to three thousand pounds per kW, significantly increasing the cost of the power train for a fuel cell powered vehicle. It is much more likely that hydrogen fuelled cars will first become affordable in the same four cylinder ICE configuration as current petrol engines.

9.5 Electrically Powered Vehicles

9.5.1 *Principle of the electric motor*

Electricity is not readily stored, and as a result, electrically powered transport has been almost entirely confined to railway or tram systems where the electricity can be provided as it is used through overhead cables or electrified rails. There have been a few battery driven vehicles, particularly for limited range use in towns, although this may change dramatically with the introduction of the hydrogen fuel cell and improved battery technology. Electric motors are already a feature of all cars, albeit

---

[d]The prototype Daimler-Benz NEBUS (New Electric BUS) utilises seven 150 litre, 900 bar roof-mounted gas bottles to store some 45,000 litres of H_2 to provide fuel for a ten stack 250 kW PEM fuel cell.

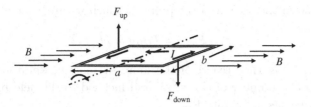

Fig. 9.9 Forces on a square loop carrying a current I in a magnetic field B.

not the prime mover: they are used as starter motors, in fans, electric windows and mirrors etc. Whether the motor is being used to drive a high speed train or just the in-vehicle CD player, the principles are the same.

A charge e moving through a uniform magnetic field $\underline{B}$ with a velocity $\underline{v}$ will experience a force:

$$\underline{F} = e\underline{v} \times \underline{B}. \qquad (9.32)^*$$

If instead, we have a length of wire carrying a current I in the magnetic field, then the wire will experience a force equal to the sum of the forces acting on each of the carriers in the wire. By definition, there will be I carriers per unit time passing some point in the wire and thus in an element $d\underline{l}$ of the wire, there will be a total of $Id\underline{l}$ carriers in motion. Substituting for $e\underline{v}$ in (9.32)

$$d\underline{F} = Id\underline{l} \times \underline{B}. \qquad (9.33)$$

The first term on the right, $Id\underline{l}$ is sometimes referred to as a current element. Integration of (9.33) allows us to calculate the force on any shape of current carrying conductor and in particular, on current loops.

Figure 9.9 shows a rectangular coil carrying a current I in uniform magnetic field $\underline{B}$. The sides length a will experience no force since they are parallel with $\underline{B}$ but the other two sides will each experience equal though opposite forces F_{up} and F_{down} of magnitude:

$$F_{\text{up}} = F_{\text{down}} = IBb. \qquad (9.34)$$

The two forces constitute a couple and will produce a torque τ that will tend to rotate the loop so that its plane is perpendicular to the direction of the magnetic field.

As the net force is zero, the torque is independent of position and we can compute the magnitude of the torque about any point that is convenient.

In the present case, it would be sensible to choose a location along one of the sides of length b, the right-hand side say, then as the perpendicular distance

---

*Note that $\underline{F}$ is perpendicular to the plane containing both $\underline{B}$ and $\underline{v}$ and is given by the right hand rule.

(lever arm) between the force and the point at which we are calculating τ is just a:

$$\tau = F_{\text{up}}a = IBab = IBA. \tag{9.35}$$

The product ab being the area A of the loop. In general, we are concerned with loops consisting of a number of turns, N, and inclined to the field by some angle θ so that the torque has a magnitude:

$$\tau = NIAB\sin\theta. \tag{9.36}$$

Torque is a vector with direction as well as magnitude and if the unit vector for the current loop is $\hat{\underline{n}}$ (determined by the right hand rule) then

$$\underline{\tau} = (NIA)\hat{\underline{n}} \times \underline{B} = \underline{m} \times \underline{B} \tag{9.37}$$

where $\underline{m}$ (Am^2) is the magnetic dipole moment. Although (9.36) and (9.37) were derived for a rectangular coil, they apply to any coil of arbitrary shape.

The torque (9.37) will turn the coil, if it is free to rotate, until the plane of the coil is perpendicular to the magnetic field direction, i.e. $\hat{\underline{n}}$ is parallel with $\underline{B}$. If by some means we arrange for the current through the coil to be reversed just as it passes through the vertical (equilibrium) position, then instead of coming to rest the coil will continue to rotate — we have in principle an electric motor. However, the coil is not stationary and its motion through the magnetic field changes things as additional currents are induced in the coil. A proper treatment of motors must include a description of the induced currents which arise from changes in the magnetic flux through the coil.

Magnetic flux ϕ is defined as the number of magnetic field lines that pass through a given area and is equal to the product of B and A. For a coil of area A inclined at an angle θ to the field (Fig. 9.10(a)) the flux is

$$\phi = \int_S N\underline{B}.\hat{\underline{n}}\,dA = NBA\cos\theta \tag{9.38}$$

where the integral is carried out over the surface, S. If the coil is rotating at constant angular velocity ω, then $\theta = \omega t$ and the instantaneous flux through the coil is:

$$\phi(t) = NAB\cos(\omega t). \tag{9.39}$$

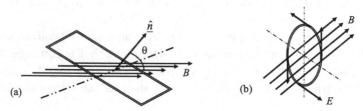

Fig. 9.10   (a) Magnetic flux through inclined coil; (b) tangential field induced in a coil by a changing magnetic field.

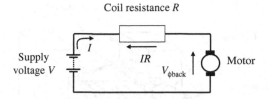

Fig. 9.11 Equivalent circuit for a DC electric motor.

Faraday's Law states that if the coil is in a changing magnetic field, an e.m.f. equal to the rate of change of the flux is induced in the coil. The e.m.f. is considered to be distributed throughout the loop and is equivalent to a tangential electric field $\underline{E}$ (Fig. 9.10(b)). The field is non-conservative, meaning that the integral around the loop is not zero and in this instance is equal to the induced e.m.f. which in turn is equal to the rate of change of magnetic flux.

$$V_\phi = \oint_C \underline{E}.d\underline{l} = -\frac{d\phi}{dt}. \tag{9.40}$$

The negative sign indicates that induced e.m.f. is always in the opposite sense to the change that generates it (Lenz's Law) and is referred to as the back e.m.f. The back e.m.f. is in effect a speed dependent impedance to the current flow.

Substituting (9.39) in (9.40), we get for the back e.m.f.

$$V_{\phi\text{back}} = -\frac{d}{dt}(NBA\cos(\omega t)) = NBA\omega\sin(\omega t). \tag{9.41}$$

We can model the performance of an electric motor using the equivalent circuit of Fig. 9.11.

Applying Kirchhoff's law to the circuit:

$$V - IR - NBA\omega\sin(\omega t) = 0 \tag{9.42}$$

where R is the resistance of the coil. Setting $\sin(\omega t) = 1$ for simplicity and rearranging, we get for the current:

$$I = \frac{V}{R} - \frac{NBA\omega}{R}. \tag{9.43}$$

The current and therefore the torque (9.37) are proportional to the angular velocity. When the power supply is first connected, the coil will accelerate until it reaches a terminal angular velocity (ω_t) when the current is just zero:

$$\omega_t = \frac{V}{NBA}. \tag{9.44}$$

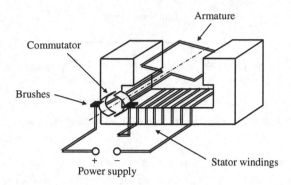

Fig. 9.12   DC electric motor with electromagnetic stator field.

A load will appear as a torque applied to the shaft in opposition to that being developed by the motor. To meet this demand, the current will increase and the speed reduce. The power delivered by the motor is therefore

$$P = V_{\phi_{\text{back}}} I = NBA\omega I = \omega\tau. \tag{9.45}$$

The torque is dependent on both the current and the magnetic field. Strictly (9.45) applies to motors employing fixed magnets. This is not always the case, and frequently the B field is provided by an electro magnet of some sort.

9.5.2 *Direct current (DC) electric motors*

The basic electric motor described in the previous section uses the torque exerted by a stationary magnetic field on a current carrying coil in which the current direction is reversed in synchrony with the rotation. The stationary field is termed the stator field and is usually provided by an electromagnet. The rotating coil assembly is known as the armature, and consists of an iron cylinder with slots cut along the surface parallel to the axis, into which the coil wires are embedded (there will generally be many coils). In DC motors, the current is supplied to the slotted coils in the correct direction and sequence through the use of *commutator assemblies*; stationary carbon brushes make contact with the relevant commutator segments and hence the coil. A simple single coil armature[e] DC motor in which the armature and stator windings are connected in series is shown in Fig. 9.12.

   The stator field windings can be connected either in series or in parallel (shunt) with the armature coils. The equivalent circuit for a series connection is shown in Fig. 9.13. The field windings are represented by an ideal coil (no resistance) in series with a total resistance, R, for the circuit and the armature windings across in which is generated the back e.m.f.

---

[e]Real motors use many armature coils.

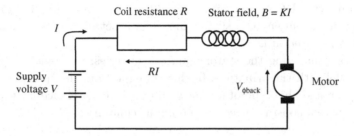

Fig. 9.13 Equivalent circuit for a DC motor a series wound stator coil.

The magnetic field developed by the stator is proportional to the current (assuming no saturation) $B = KI$ where K is some constant. If the inductance of the stator field windings is sufficiently low, then the potential across the stator becomes small in relation to the other potentials in the circuit.[f] With this simplification, the circuit is effectively identical to that in Fig. 9.12, except that the stator field is now a function of the current in the circuit.

Substituting for B in (9.42) and setting $\sin(\omega t) = 1$ as before:

$$V - IR - NA(KI)\omega = 0 \qquad (9.46)$$

and in (9.45):

$$P = \tau\omega = I^2KNA\omega : \quad \text{or} \quad \tau = I^2KNA. \qquad (9.47)$$

From (9.47), we get $I = \sqrt{\tau/KNA}$ and substituting this into (9.46) and rearranging the equation we obtain the torque-rotational speed characteristic for the motor:

$$\tau = \frac{V^2(KNA)}{(KNA\omega + R)^2}. \qquad (9.48)$$

Clearly, in this type of motor, the torque is essentially constant and close to maximum at low speeds ($KNA\omega << R$), but above this the torque starts to decrease rapidly as the square of the speed:

$$\tau \to \frac{V^2(KNA)}{R^2} : \quad \omega \ll \frac{R}{KNA} \qquad (9.49)$$

and

$$\tau \to \frac{V^2}{KNA\omega^2} : \quad \omega \gg \frac{R}{KNA}. \qquad (9.50)$$

---

[f]This is a gross assumption for the single coil armature we are discussing here. In real motors with several armature coils the approximation is much better.

The series connected DC motor is best suited for applications requiring high torque at low speed, for example, as a starter motor in a petrol or diesel engine where a high torque is required at $\omega = 0$.

Instead of connecting the stator windings in series, we could have connected them in parallel with the armature. In this configuration, the speed varies linearly with the torque and this type of motor would be a better choice for the in-vehicle CD player, where a constant speed is the main requirement.

9.5.3 *Alternating current (AC) electric motors: Induction motors*

In the series connected DC motor, the same current flows through both the stator and the armature coils. We might therefore suppose that such a motor would run equally well from an AC source (e.g. 50 Hz mains electrical supply). However, this is to ignore the inductance (L) of the field (stator) and armature windings. When connected to an AC supply, these coils present an increased impedance ($i\omega L$), thus limiting the current, and introducing a phase difference between the field and armature currents (inductive impedance is reactive).

It is possible to use low inductance armatures with AC supplies (referred to as universal motors), but these tend to have low power, low torque characteristics, suitable for use in, for example, domestic appliances.

The most common AC motor is the induction motor. This dispenses with the commutator altogether, replacing it with shorting plates, which as the name suggests, short circuits the ends of the armature coils. The AC supply is only connected to the stator windings and through the judicious arrangement of the stator windings (which remain stationary), the *stator field* is made to rotate generating an e.m.f. across the rotor conductors, driving current through them. The rotating stator field produces a torque on the current in the armature windings causing the armature to turn at a speed that will tend to minimise the induced e.m.f. and hence the current.

When no external load is placed on the motor, it will rotate at virtually the same speed as the stator field. When the motor is loaded, its speed will start to decrease. There is now a significant e.m.f. induced in the armature windings at a frequency equal to the difference between the rotational speeds of the stator field and the armature. The rotor inductance results in an impedance proportional to the speed difference, causing the current to lag the voltage. Maximum torque will occur when the phase difference is just sufficient to prevent the proper interaction between the armature current and stator field. Increasing the load further would cause the motor to stall (Fig. 9.14).

A particular advantage of AC induction motors, is that no commutator assemblies are required. In high power applications, the brushes tend to wear out rapidly increasing the maintenance requirements. A disadvantage is the need for an AC supply, a significant impediment to their use in hydrogen fuel cells, since they generate DC. This has led to the development of DC based induction motors,

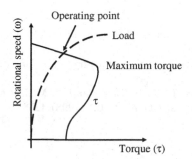

Fig. 9.14   Torque-speed characteristic for an AC induction motor.

where sophisticated high power electronic systems are used to control the stator fields. With routine use of engine management systems in road vehicles, this is less of a problem now than would have been the case ten to fifteen years ago.

9.6 Problems

1. An internal combustion engine with a compression ratio of 8 is designed to burn octane (C_8H_{18}) in air. Using the data in Table 2.1, determine the fuel: ratio required for complete combustion of the fuel and calculate the specific gas constant for the mixture. If the mean value of c_p over the cycle is $1.41 \, \text{kJ} \, \text{kg}^{-1} \, \text{K}^{-1}$, find the value of γ and hence the ideal efficiency.

2. If in the engine in question, the fuel/air mixture is admitted at atmospheric pressure ($100 \, \text{kPa}$) and at a temperature of $340 \, \text{K}$, find the temperature and pressure at the end of compression.

3. A $5\frac{1}{2} \, \text{kW}$, $110 \, \text{V}$ power system is to be constructed from an assembly of H_2 fuel cells.

   Determine:

   (a) the voltage produced by an ideal single cell;
   (b) the number of such cells and the connection required to realise the supply;
   (c) the rate at which H_2 would be consumed by the supply, in mole s^{-1};
   (d) the theoretical maximum efficiency.

   Comment on what other factors would contribute to losses in the system.

   (Assume: Gibb's free energy (ΔG) for liquid water $= -241 \, \text{kJ} \, \text{mole}^{-1}$;
   Heat of formation of water (enthalpy, ΔH_f) $= -286 \, \text{kJ} \, \text{mole}^{-1}$.)

4. In a shunt (parallel) wound DC electric motor, the stator field is, in principle, not dependent on the armature current. Show that the speed in such a motor is linear in the torque. What is the speed of the motor when there is no torque demand?
   (Hint: draw the equivalent circuit and apply Kirchhoff's voltage law.)

References

1. P. Hoffmann, *Tomorrow's Energy. Hydrogen, Fuel Cells and the Prospects for a Cleaner Planet*, (MIT Press, 2001).
2. Sir William Grove, On the gas voltaic battery, *Philosophical Magazine and Journal of Science*, 1843, p. 272.
3. A. C. Chambers, C. Park, R. T. K. Baker and N. M. Rodriguez, Hydrogen storage in graphite nanofibres, *J. Phys. Chem.* B **102** (1998) 4253.

Chapter 10

TRANSPORT AND DISPERSAL OF POLLUTANTS IN THE ENVIRONMENT

10.1 Introduction

All waste products that are inimical to life have ultimately to be accommodated within the environment, and the proper dispersal of pollutants is therefore critically important if life is to continue on the planet. Usually, this involves dilution of the contaminating species to some safe level in rivers or the atmosphere, or else burial in landfill. In nature, a variety of biological processes have evolved over millennia to ensure the balance of life is maintained. Occasionally, the balance is disturbed, for example by major volcanic eruptions which eject large volumes of toxic matter into the atmosphere with drastic consequences for the biosphere. The increasing levels of human activity are also making significant impact on the environment. We have already discussed the effects of CO_2 and CFCs on global warming and ozone depletion.

In this chapter, we shall consider the processes involved in the transportation of low concentrations of pollutants, sometimes referred to as guest species, through some host medium such as water, the ground or the air. The underlying physics is quite general and applies irrespective of the nature of the guest specie/contaminant or the host. The magnitudes of controlling parameters will vary strongly with the situation, but the basic mechanisms of diffusion, advection and turbulence are common to many of the transport process of concern here.

Consequently, we shall first consider the underlying physical principles and apply them in a few generic situations, before proceeding to a discussion of some specific cases of relevance including: dispersal in rivers[1] and smoke from chimneys, where we shall consider the role of turbulence in dilution and ground water flow.

10.2 Transport Physics: Diffusion and Advection

Diffusion is a direct consequence of entirely random collisions between the atoms and molecules of the guest species and those of the host medium (i.e. air, water etc.). Individual collision or scattering events are independent and the speeds and direction of the scattered particles are arbitrary. If the concentration of scattering centres is high, the mean distance between collisions (mean free path) will be low and vice versa. If the distribution of such centres is uniform, there will be no net motion of the scattered species. Conversely, if there is a variation along some direction

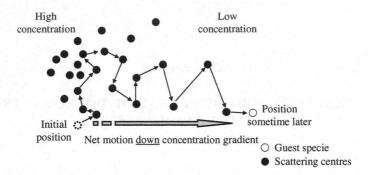

Fig. 10.1   Diffusion of a guest particle down a concentration gradient.

(i.e. a concentration gradient), the mean free path will vary, it will be greater where the concentration is lower. On the average therefore, particles will travel farther in the direction of low concentration and there will be a net transport — diffusion — *down* the concentration gradient (Fig. 10.1). The flux $\underline{F}$ (mass diffusing through unit area per unit time) is directly proportional to the concentration gradient

$$\underline{F} \propto -\nabla C(\underline{r},t) = -D\nabla C(\underline{r},t) \tag{10.1}$$

where $C(\underline{r},t)$ is the particle concentration (mass per unit volume) and the constant of proportionality D is the Diffusion Coefficient ($m^2 s^{-1}$) which is the area swept out by the diffusing species in unit time. Equation (10.1) is Fick's law of diffusion, the negative sign indicating that the flux is directed down the gradient. It is assumed that the concentration is low, by which we mean that the sample volume is large in relation to the mean separation between particles and that the change in mass of volume elements due to diffusion is negligibly small.

The concentration is a function both of space $(\underline{r})$[a] and time (t) and in any given sample volume, the concentration and distribution of the guest species will therefore, in general, vary as a result of diffusion. If we consider a (fixed) sample volume V bounded by a closed surface S_c then the outflow of particles, i.e. $\underline{F}$, through S_c from V is related by the Gauss divergence theorem

$$\int_{Sc} \underline{F}.d\underline{s} = \int_V \nabla.\underline{F}\,dV. \tag{10.2}$$

The total mass of particles in the sample volume must be the integral of C over V. If in time dt there is an outflow of particles, then (assuming no sinks or sources) the change in concentration will be $-dC$. The total reduction in the mass of particles in the volume must equal the flux that has flowed out through the surface, and this

---

[a] $|r| = \sqrt{x^2 + y^2 + z^2}$.

is given by the product of dt and (10.2) and in particular the right-hand side of the equation:

$$\int_V (-dC)dV = \int_V \nabla \cdot \underline{F} dV dt. \tag{10.3}$$

Since this is true for all volumes and times, the integrands must be equal, in particular:

$$-dC = \nabla \cdot \underline{F} dt. \tag{10.4}$$

Rearranging (10.4) we obtain the equation of continuity or conservation of mass (i.e. flux in or out must be equal to the corresponding increase or reduction in concentration):

$$\frac{\partial C}{\partial t} + \nabla \cdot \underline{F} = 0. \tag{10.5}$$

Note that we had specified a fixed volume, V and so we have to use a partial derivative with respect to time for the concentration as the space coordinates were all fixed.

Substituting for the flux using Fick's law (10.1), we get the diffusion equation for a stationary host medium:

$$\frac{\partial C}{\partial t} + \nabla \cdot (-D\nabla C) = \frac{\partial C}{\partial t} - D\nabla^2 C = 0. \tag{10.6}$$

Equations of the form (10.6) occur often in apparently widely differing circumstances; for example it describes the diffusion of heat through a conducting medium, neutrons through a uniform reactor core, in short wherever a diffusion-like process is taking place.

Invariably, the host medium is moving (e.g. a river) with some velocity $\underline{u}$ and the guest particles will become entrained in the flow. This is termed advection and in the absence of diffusion, the flux would be given by

$$\underline{F} = \underline{u}C. \tag{10.7}$$

The total flux is just the sum of the diffusing and advected components.

$$\underline{F} = \underline{u}C - D\nabla C. \tag{10.8}$$

Substituting this into the continuity equation (10.5)

$$\frac{\partial C}{\partial t} + \nabla \cdot \{\underline{u}C - D\nabla C\} = \frac{\partial C}{\partial t} + C\nabla \cdot \underline{u} + \underline{u}.\nabla C - D\nabla^2 C = 0. \tag{10.9}$$

If we make the simplifying assumption that $\nabla \cdot \underline{u} = 0$ (which it should be noted is only a reasonable approximation in simple cases), then (10.9) reduces to

$$\frac{\partial C}{\partial t} + \underline{u} \cdot \nabla C - D\nabla^2 C = 0. \tag{10.10}$$

Replacing the first two terms on the left hand side of (10.10) using the operator identity[b]

$$\frac{d}{dt} = \frac{\partial}{\partial t} + \underline{u} \cdot \nabla \tag{10.11}$$

we get the Transport Equation for the guest species in the presence of both diffusion and advection

$$\frac{dC}{dt} = D\nabla^2 C. \tag{10.12}$$

10.3 Generic Examples

10.3.1 *Instantaneous point source in a stationary medium*

A common class of problems is the instantaneous release of some guest particles into a host medium at rest ($\underline{u} = 0$). Although this is a slightly artificial situation in that host media are rarely stationary, the analysis of this type of problem provides some insights into more realistic situations, which may not be amenable to easy solution.

Consider the release of mass M of guest particles into a stationary, isotropic and homogenous medium at time $t = 0$ at the origin ($\underline{r} = 0$). We would expect that although the trajectory of any particular particle would be random, the guest species taken as a whole would diffuse away from the origin over time in a statistically uniform manner. There are no preferred directions and the concentration should decay with distance.

The solution to (10.12) is the Gaussian function

$$C(\underline{r}, t) = \frac{M}{(\sigma\sqrt{2\pi})^n} \exp\left\{\frac{-r^2}{2\sigma^2}\right\} \quad \text{where the standard deviation } \sigma^2 = 2Dt.$$
$$\tag{10.13}$$

The index (n) depends on the symmetry of the situation. In the present case, there is full spherical symmetry and $n = 3$. Equally for line sources where there is axial symmetry $n = 2$ and for a plane source with just one principle direction, $n = 1$.

With time, the guest particles will disperse over a spherical volume, with the concentration reducing monotonically at the origin (Fig. 10.2). The distribution broadens and reduces in magnitude as a result, although the total mass of particles remains constant.

---

[b]Consider the total derivative of the function $f(x(t), y(t), z(t), t)$ $\frac{df}{dt} = \frac{\partial f}{\partial t} + \frac{\partial f}{\partial x}\frac{dx}{dt} + \frac{\partial f}{\partial y}\frac{dy}{dt} + \frac{\partial f}{\partial z}\frac{dz}{dt}$ and as $\frac{dx}{dt}, \frac{dy}{dt}, \frac{dz}{dt}$ are the x, y, and z components of the velocity $\underline{u}$, we get $\frac{df}{dt} = \frac{\partial f}{\partial t} + \nabla f.\underline{u} = \{\frac{\partial}{\partial t} + \underline{u}.\nabla\}f$. The term $(\underline{u}. \nabla)$ is the 'Advective Operator'.

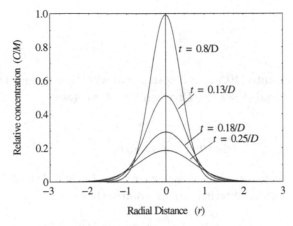

Fig. 10.2   Radial distribution of particles at different times after an instantaneous release into a stationary medium.

10.3.2 *Continuous point source*

A second illustrative type of problem is the continuous release of a guest species from a point source into a stationary homogeneous, isotropic medium. Again, this is a somewhat artificial state of affairs, but builds on the previous result.

Suppose we have such a continuous source located at the origin that starts emitting at a rate m $(\mathrm{kg\,s^{-1}})$ at time $t = 0$. We wish to determine the concentration C at some position $\underline{r}$ at a time t later. In the first elemental interval of time $d\tau$ the source will have emitted an amount $md\tau$ which will diffuse to $\underline{r}$ in due course. The concentration due to this (effectively instantaneous source) will be given by (10.13). However the source is a continuous one and during the next interval $d\tau$ there will be another quantity $md\tau$ released which will also make some contribution to the concentration at $\underline{r}$. The second component was emitted at time τ later and therefore has had less time to diffuse to $\underline{r}$, i.e. (t-τ). To sum all the elemental contributions up to time t, we need to integrate (10.13) with respect to τ and remember to include the cumulative effect of all the delays, by noting that in this case, $\sigma^2 = 2D(t - \tau)$.

$$C(\underline{r}) = \int_0^t \frac{m}{8(\pi D)^{3/2}(t - \tau)^{3/2}} \exp\left(-\frac{r^2}{4D(t - \tau)}\right) d\tau. \qquad (10.14)^{\mathrm{c}}$$

In dealing with an integral such as this, it is best to effect a change of variable of the form:

$$\beta^2 = \frac{r^2}{4D(t - \tau)}. \qquad (10.15)$$

---

[c]This type of operation is termed convolution and (10.14) is a convolution integral.

It follows that:

$$d\tau = \frac{4\sqrt{D}}{r}(t-\tau)^{3/2}d\beta. \qquad (10.16)$$

Substituting for this into (10.14) and noting that with the change of variable (10.15), $\beta = \infty$ when $\tau = t$ and $\beta = r/2\sqrt{Dt}$ when $\tau = 0$, we get

$$C(\underline{r}) = \frac{m}{4\pi Dr}\left\{\frac{2}{\sqrt{\pi}}\int_{r/2\sqrt{Dt}}^{\infty}\exp(-\beta^2)d\beta\right\}. \qquad (10.17)$$

The term in the curly brackets is known as the Complementary Error Function (erfc)[d] and (10.17) may be written more compactly as:

$$C(\underline{r}) = \frac{m}{4\pi Dr}\mathrm{erfc}(r/2\sqrt{Dt}). \qquad (10.18)$$

When the argument $(r/2\sqrt{Dt})$ of the complementary error function is large (i.e. at short times and/or large radial distances), its value and hence the particle concentration $\to 0$. Conversely, at long times and/or short distances the $\mathrm{erfc}(r/2\sqrt{Dt}) \sim 1$, and the concentration varies approximately as $1/r$.

10.3.3 *Continuous point source in a flowing medium*

The final representative example is to consider the distribution of the guest particles species being emitted from a continuous source into a medium flowing with a constant velocity, $\underline{u}$. As before, we assume that the source starts emitting at time $t = 0$, but in this case we have the added complication that the medium is moving, though the source is fixed. By convention, we align the velocity along the x-axis. With the source at the origin we choose a coordinate system, (x', y', z') moving with the fluid and use a Galilean transformation to return to the stationary frame, i.e.

$$x' = x - ut; \quad y' = y; \quad z' = z. \qquad (10.19)$$

In essence, we follow the same procedure as for the continuous emission into a medium, except that we must substitute:

$$(r')^2 = (x - u(t-\tau))^2 + y^2 + z^2 \qquad (10.20)$$

in the argument of the exponential. We substitute (10.20) into (10.14) to get

$$C = \frac{m}{8(\pi D)^{3/2}}\int_{r/2\sqrt{Dt}}^{\infty}\exp\left(-\frac{(r')^2}{4D(t-\tau)}\right)\frac{d\tau}{(t-\tau)^{3/2}}. \qquad (10.21)$$

---

[d]The complementary error function is defined as $(1 - \mathrm{erf}(x))$ where $\mathrm{erf}(x) = \frac{2}{\sqrt{\pi}}\int_0^x \exp(t)dt$ is the error function. Values of $\mathrm{erf}(x)$ and $\mathrm{erfc}(x)$ are obtained from standard tables, e.g. M. R. Spiegel and J. Liu, *Mathematical Handbook of Formulas and Tables*, (McGraw Hill, 1999).

For very long times ($t \to \infty$), the lower limit of integration ($\propto (1/\sqrt{t})$) becomes effectively zero. Evaluating the integral therefore from $t = 0$ to ∞ with the help of integration tables:

$$C = \frac{m}{4\pi r D} \exp\left(-\frac{u(r-x)}{2D}\right). \tag{10.22}$$

In the absence of turbulence (10.22) gives the steady state concentration at some point r. A long way downstream, i.e. large x, $x^2 \gg (y^2 + z^2)$ and using the binomial approximation:

$$r = \sqrt{x^2(1 + (y^2 + z^2)/x^2)} \approx x + (y^2 + z^2)/2x \tag{10.23}$$

and substituting in (10.22):

$$C = \frac{m}{4\pi r D} \exp\left(-\frac{u(y^2 + z^2)}{4Dx}\right). \tag{10.24}$$

Equation (10.24) maps the radial concentration profiles about the flow direction, and shows that these decay exponentially with radial distance, except near the axis of flow (y, z small) when the concentration falls off as $(1/r)$.

The ratio (x/u) may be interpreted as the time that has elapsed since a given element, $md\tau$ was emitted by the source. The standard deviation (10.13) can then be written

$$\sigma^2 = 2Dt \approx 2Dx/u. \tag{10.25}$$

10.4 Dispersal in Turbulent Rivers

Flow in real rivers is turbulent and unstable, and as a result dispersion patterns will vary from moment to moment. The process is statistical and not easy to describe analytically. We have to rewrite equation (10.8) for the flux therefore, in terms of effective diffusion constants that describe in some sense the average behaviour of the process.

$$\underline{F} = \underline{u}C + \left\{ \varepsilon_x \frac{\partial C}{\partial x}\hat{\underline{i}} + \varepsilon_y \frac{\partial C}{\partial y}\hat{\underline{j}} + \varepsilon_z \frac{\partial C}{\partial z}\hat{\underline{k}} \right\} \tag{10.26}$$

where ε_x, ε_y and ε_z are the Coefficients of Turbulent Diffusion in the x, y and z directions respectively.

We shall take as a starting point the continuous emission from a point source in a river from, say, a waste treatment plant at the origin ($\underline{r} = 0$). Figure 10.3 shows a section through some river, of width W, where the depth varies as $h(y)$ and the downstream velocity as $\underline{u}(y)$. Although increasingly implausible, we shall continue to assume for the sake of simplicity that $\nabla \cdot \underline{u} = 0$.

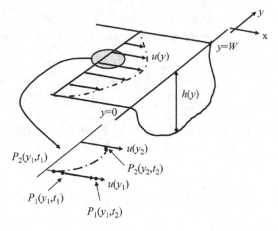

Fig. 10.3   Cross section through a river.

If the river is shallow in relation to its width (as is usually the case), vertical mixing ($\partial C/\partial z = 0$) will as a rule be rapid and for practical purposes, the source may be considered as a two dimensional line source of strength ($m/h(0)\,\mathrm{kg\,m^{-1}\,s^{-1}}$).

Now suppose a molecule located at y_1 (see Fig. 10.3) with velocity $u(y_1)$ diffuses some distance in the y-direction to a new position y_2. The molecule will in due course be entrained (advected) into the flow at y_2 and acquire a velocity $u(y_2)$, which in general will not be the same as that it initially had at y_1. Consequently, it will be increasingly separated or dispersed from molecules remaining at y_1 due to the dissimilar downstream velocities. Dispersion therefore is a consequence of the interaction between transverse (y-direction) diffusion and longitudinal or downstream (x-direction) advection.

We select a small volume of water of width dy, length dx and unit depth — since we are considering the case where vertical mixing has already been completed — as it flows down the river (Fig. 10.4). There will be continuous inflows and outflows of molecules along both the transverse y direction, i.e. at y and $y + dy$ and the longitudinal x direction, x and $x + dx$. The difference in the y inflows and outflows must be accounted for by a change in the mean concentration within the volume

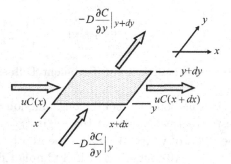

Fig. 10.4   Inflow and outflow of guest species in a volume element moving with constant velocity, u.

element. Molecular movement in y is diffusive, while in x it is primarily advective and proportional to the mean velocity and the concentration (10.7). Consequently, the advection of mass down stream will depend on the mass diffusing in laterally and although advection separates particles along the stream direction, it is transverse diffusion that is important for mixing. For the sake of simplicity, we shall neglect turbulent diffusion along the x-direction and will only consider ε_y, provided we are sufficiently far downstream so that vertical mixing has been completed.

The concentration is obtained from (10.24), by replacing D by ε_y, r in the pre-exponential term[e] by x and using the mean velocity, $\bar{u}$ defined as:

$$\bar{u} = \frac{\int_0^W u(y)h(y)dy}{\int_0^W h(y)dy} \tag{10.27}$$

where the term in the denominator is the cross-sectional area, to get:

$$C = \left(\frac{m}{h}\right) \frac{1}{\sqrt{4\pi\varepsilon_y x\bar{u}}} \exp\left\{-\frac{\bar{u}y^2}{4\varepsilon_y x}\right\}. \tag{10.28}[f]$$

Strictly speaking, (10.28) only applies to a very wide river, where any effects due to the river banks are small and may be neglected. In the more general situation, we need to include the effects of the river boundaries. Consider the case where the source (i.e. the origin) is situated at the centre line of a straight river (or perhaps a canal) of width W, the banks of which will therefore be at $y = \pm\frac{1}{2}W$. A not unreasonable boundary condition would be to assume that the net flux at the edge of the river is zero; i.e. in the steady state, pollutants being absorbed is equal to that being released back into the river. Algebraically, this would be expressed, using the one dimensional form of Fick's law as:

$$F = -\frac{\partial C}{\partial y}\bigg|_{\pm\frac{W}{2}} = 0. \tag{10.29}$$

Taking the bank at $y = \frac{1}{2}W$, for example, and differentiating (10.28) with respect to y, we find that, except at very large x (a long way downstream) or y (a very wide river):

$$F = -\frac{d}{dy}\left\{\frac{m}{h\sqrt{4\pi\varepsilon_y x\bar{u}}} \exp\left[-\frac{\bar{u}y^2}{4\varepsilon_y x}\right]\right\} = \frac{m\sqrt{\pi\bar{u}}}{h(4\pi\varepsilon_y x)^{\frac{3}{2}}} W \exp\left[-\frac{\bar{u}W^2}{16\varepsilon_y x}\right] \neq 0.$$
$$\tag{10.30}$$

In other words the boundary condition cannot be fulfilled; similarly at the other bank.

---

[e]From (10.23), $r \sim x$ for $x \gg y^2/2x$; note we have already assumed the river is shallow in relation to its width and length.

[f]For a line source, the index in (10.13) is $n = 2$, as opposed to a point source where $n = 3$. Consequently (10.24) must be multiplied by $\sqrt{2\pi\sigma^2}$ using (10.25) where $\sigma^2 = 2\varepsilon_y(x/\bar{u})$.

Mathematically, we can add a *virtual source* of the same strength at an equal distance beyond the river edge (i.e. at $y = W$) to cancel the flux at $y = \frac{1}{2}W$ (i.e. force the boundary condition to be fulfilled). However, this would result in an even greater flux at the other bank, $y = -\frac{1}{2}W$. Adding a similar virtual source on that side would simply increase the flux at the first bank $(+\frac{1}{2}W)$ and so on. In fact we need a double infinite series of "mirror sources", as they are termed, at $y = \pm W, \pm 2W, \pm 3W \ldots$ to properly describe the mixing process in a river with finite banks.

$$C = \frac{m}{h\sqrt{4\pi\varepsilon_y x\overline{u}}} \sum_{n=0}^{\infty} \left\{ \exp\left[-\frac{\overline{u}(y-nW)^2}{4\varepsilon_y x}\right] + \exp\left[-\frac{\overline{u}(y+nW)^2}{4\varepsilon_y x}\right] \right\}. \quad (10.31)$$

If the actual source is not located at the centre line of the river then the distance from the source to the centre line must be added and subtracted from the origin of the mirror sources accordingly. Physically, y must lie between the river banks, that is between $\pm\frac{1}{2}W$. Inserting the maximum positive value for y into (10.31), and taking out common factors:

$$C = \frac{m}{h\sqrt{4\pi\varepsilon_y x\overline{u}}} \sum_{0}^{\infty} \exp\left\{-\frac{\overline{u}W^2}{4\varepsilon_y x}\right\} \left\{ \exp\left[\left(\frac{1}{2}-n\right)^2\right] + \exp\left[\left(\frac{1}{2}+n\right)^2\right] \right\}. \quad (10.32)$$

The term $\overline{u}W^2/\varepsilon_y$ has units of length. It provides a distance scale for mixing of pollutants in the river. Empirically it is found that about one tenth of this length is required to achieve complete ($> 95\%$) mixing across the river, a distance not surprisingly known as the mixing length:

$$L_m = 0.1 \times \frac{\overline{u}W^2}{\varepsilon_y}. \quad (10.33)$$

The mixing length increases with the speed of the river; in the time it takes for mixing to occur, the contaminant will have moved further downstream if the river is flowing more rapidly. Similarly we would expect mixing to take place more quickly when the coefficient of turbulent diffusion is greater. Less obvious is the dependence on W^2; for mixing to be complete, the species must diffuse transversely across the width of the river and if the river is wide, this will take longer.

Most turbulence in rivers arises as a result of non-uniformity in the river profile. The banks are not smooth and are commonly overgrown with vegetation; there will be stones on the river bed etc. However, even in a smooth straight concrete channel of constant depth, turbulence will still be the dominant process due to friction effects.

The pressure acting on the floor of the river bed due to the water above it is given in the usual way as $(\rho g h)$ where ρ is the density, and the component of this along the downhill slope will provide the force causing the river to flow. There will

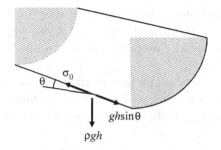

Fig. 10.5   Frictional forces along stream bed.

also be a frictional stress (σ_0) along the river bed opposing the motion and if the river is flowing with constant speed, then the forces must be in equilibrium:

$$\sigma_0 = \rho g h \sin \theta \approx \rho g h \theta \qquad (10.34)$$

where θ is the slope of the river bed (Fig. 10.5). If we divide both sides of (10.34) by the density, we end up with a quantity that has units of $(\mathrm{ms}^{-1})^2$, i.e. velocity squared. The situation is directly analogous to that discussed in Sec. 8.4.3 for the vertical dependence of wind speed. In that case, we defined a shear or friction velocity, u_* associated with a boundary layer. Here the friction velocity is defined by

$$u_* = \sqrt{\frac{\sigma_0}{\rho}} = \sqrt{g h \theta}. \qquad (10.35)$$

Turbulence will be associated with the behaviour of the boundary layer and we note that the coefficients of turbulent diffusion (ε_x, ε_y and ε_z) have units of $\mathrm{m}^2\mathrm{s}^{-1}$, area swept per unit time. It is useful to describe these parameters in terms of the friction velocity and the river depth.

$$\varepsilon = \kappa u_* h \qquad (10.36)$$

where κ is a correction factor, intended to account for different types of flow. A smooth straight channel will have a small value of κ (~ 0.2), while a meandering stream might be expected to have a larger value (> 0.5). For ε_z, κ is usually taken to be 0.067.

10.5 Emission of Smoke from Chimneys

The visible smoke we observe being emitted from smoke stacks at power stations and vehicle exhausts (Sec. 9.3.3) for example, is a mixture of small particles (aerosols) and gases such as CO_2. The situation is in many ways similar to the release of effluent in rivers: there is a predominant direction of flow, i.e. prevailing wind direction, transverse diffusion is limiting, but unlike rivers there is no finite well-defined upper boundary (equivalent to the surface of the river) and hence vertical (z) mixing will take place on time and distance scales comparable to the transverse (y) direction. In addition, there are convection and buoyancy effects which act primarily in the vertical direction. Consequently, we cannot assume a line source for the emitter as we did in the case of (10.28).

Mixing is turbulent — as is the case in rivers — and equation (10.26) serves as a useful starting point with the proviso that the coefficient of turbulent diffusion in the vertical direction (ε_z) includes a stronger buoyancy component. Including conservation of mass:

$$\frac{\partial C}{\partial t} + u\frac{\partial C}{\partial x} = \left\{ \varepsilon_x \frac{\partial^2 C}{\partial x^2} + \varepsilon_y \frac{\partial^2 C}{\partial y^2} + \varepsilon_z \frac{\partial^2 C}{\partial z^2} \right\}. \tag{10.37}$$

Consider a chimney at $x = y = 0$ of height $z = h$ that is emitting m (kg s^{-1}) of smoke into a uniform wind of velocity u blowing along the x direction. In addition we assume that there is total reflection of smoke particles from ground level ($z = 0$), which we deal with mathematically by adding a fictitious source of equal magnitude at $z = -h$ (since there is no upper boundary, we do not have to add multiple sources). The concentration downwind of the chimney can then be shown to be:

$$C = \frac{m}{2\pi u \sigma_y \sigma_z} \left\{ \exp\left(-\frac{y^2}{2\sigma_y^2} - \frac{(z-h)^2}{2\sigma_z^2} \right) + \exp\left(-\frac{y^2}{2\sigma_y^2} - \frac{(z+h)^2}{2\sigma_z^2} \right) \right\} \tag{10.38}$$

where σ_y and σ_z[g] the corresponding standard deviations in the y and z directions.

We are, of course, most concerned about smoke at the ground, $z = 0$, and this will be greatest along the $y = 0$ axis. Equation (10.38) then reduces to:

$$C = \frac{m}{\pi u \sigma_y \sigma_z} \left\{ \exp\left(-\frac{h^2}{2\sigma_z^2} \right) \right\}. \tag{10.39}$$

The heavier particulate emissions will settle out of the smoke quite rapidly, but the lighter fractions and the gaseous components can remain in the atmosphere for long periods of time during which they may be transported round the globe and even up through the troposphere into the stratosphere.

The standard deviations σ_y^2 and σ_z^2 correspond to the mean squares of the smoke profiles $\langle y^2 \rangle$ and $\langle z^2 \rangle$ respectively. The value of $\langle z^2 \rangle$ after time t is

$$\sigma_z = \sqrt{\langle z^2 \rangle} t = u_{z\mathrm{RMS}} t. \tag{10.40}$$

The root mean square (RMS) vertical component of the velocity (u_{zRMS}) will depend partly on the amount of turbulence and partly on the rate of convection. Assuming that the smoke being emitted from the chimney is warmer than the surroundings, convective forces will cause it to rise. In stable or neutral atmospheres (Sec. 2.3.2), where the lapse rate of the smoke is the same as that of the environment, vertical mixing will be slow, therefore u_{zRMS} and hence σ_z will be small. The plume

---

[g] $\sigma_y^2 = 2\varepsilon_y t$, & $\sigma_z^2 = 2\varepsilon_z t$; $t = u/x$.

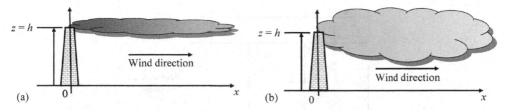

Fig. 10.6 Emission of smoke into atmospheres that are (a) stable; (b) unstable.

of smoke will remain narrow, and extend a long way downwind of the chimney at roughly the same height above the ground (Fig. 10.6(a)). In an unstable atmosphere, where the lapse rate is greater than that of the surroundings, buoyancy forces will cause greater mixing and turbulence and the smoke will disperse more rapidly (Fig. 10.6b).

Smoke will be emitted into the boundary layer of the atmosphere where turbulence is predominant and the wind speed relatively low. Consequently, at low levels, mixing is comparatively rapid, and smoke plumes are dispersed over short distances. At higher altitudes in the boundary layer, the wind speed (u) increases logarithmically as in (8.24) and advection effects become more important. Smoke particles that diffuse above the boundary layer can be transported great distances by the prevailing wind systems. Acid rain falling in Norway and Sweden has been traced to sulphur emission from coal-fired power stations in Britain carried on the westerly winds, and radioactive contamination from Chernobyl was deposited by rainfall over parts of the British Isles.

10.6 Ground Water Flow

10.6.1 *Darcy's Equation*

Soils are composed of grains of various size, compacted together with pores in between through which water may flow. In dense soils, for example clays, the level of compaction is high and as a result the pore sizes are small. It is difficult for water to percolate through such soils and they are more or less impervious to the flow of water. Conversely, sand is composed of rather coarse loosely packed grains with large pores through which water can flow relatively freely.

The flow of water through the ground was first studied by the French water engineer, Henri Darcy while designing the water supply for Dijon in 1856 and he first derived the relationship between hydraulic potential (the capacity to 'drive' water through the pores) and hydraulic conductivity (i.e. ease with which water may flow through a given type of soil).

A schematic cross section though the top soil is shown in Fig. 10.7. In general, the upper levels will not be saturated as rain water will have percolated down through the pores under gravity. At some depth all the available pore spaces will be filled with water and the soil becomes saturated. The boundary between the

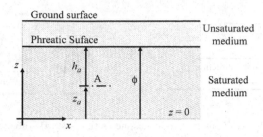

Fig. 10.7   Section through a water bearing medium.

saturated and unsaturated zones is known as the 'Phreatic Surface', and as it is in open contact with the atmosphere, the pressure on the water in the pores will be just equal to atmospheric pressure. If a well were to be sunk, the level of water in the well would correspond to the phreatic surface. Above it the pore pressure will be negative with respect to atmospheric pressure as a result of capillary action. In the saturated zone, the water in the pores will be under hydrostatic (positive) pressure given by the height of the column of water measured to the phreatic surface.

Take the point A in Fig. 10.7, the pore pressure here is:

$$p_a = h_a \rho g \qquad (10.41)$$

where h_a is the 'head', the depth of A below the phreatic surface. We select an arbitrary datum level in the saturated zone, ($z = 0$), and define the head, h above this to be the 'Hydraulic Potential', ϕ, which will be given by the sum of z_a and h_a in Fig. 10.7 or generalising:

$$\phi = h + z \quad \text{or} \quad h = \phi - z. \qquad (10.42)$$

If there is a pressure gradient there will be a corresponding flow of water through the pores down the gradient. The volume flow through unit area in unit time ($m^3/m^2/s \equiv ms^{-1}$) is the 'Specific Discharge Vector', q. Along the x-direction the discharge is, using (10.41) and (10.42):

$$q_x \propto -\frac{\partial p}{\partial x} = -\frac{\rho g k_p}{\mu_d}\frac{\partial h}{\partial x} = -K\frac{\partial \phi}{\partial x} \qquad (10.43a)$$

where μ_d is the dynamic viscosity of the water and k_p the specific or intrinsic permeability of the medium. The constant of proportionality, K is the 'Hydraulic Conductivity', and the minus sign signifies flow is down the pressure gradient.

Similarly, the y-component of the discharge is:

$$q_y = -K\frac{\partial \phi}{\partial y}. \qquad (10.43b)$$

For discharge in the vertical direction, we must include gravity in (10.43):

$$q_z = -\frac{\rho g k_p}{\mu_d}\frac{\partial p}{\partial z} - \frac{k_p}{\mu_d}(\rho g) = -K\frac{\partial}{\partial z}(\phi - z) - K = -K\frac{\partial \phi}{\partial z}. \qquad (10.43c)$$

Expressing (10.43a–10.43c) in three dimensions:

$$\underline{q} = -K\nabla\phi \tag{10.44}$$

which is the usual form of Darcy's equation.

Apart from g, there are three principal contributions to the hydraulic conductivity: two of them — the viscosity and the density — are properties of the water and the third — the specific permeability — is a property of the medium. We have assumed implicitly in the derivation of Darcy's equation that these were all isotropic, homogenous parameters. In practice this is unlikely. The density, for example, will vary with the salinity and temperature and the soil will normally be anisotropic; in general, the conductivity will therefore be a tensor. While the density, viscosity and permeability are seldom isotropic, they are frequently slowly changing variables and may be considered as effectively constant in time and space.

Discharge rate was defined as volume flowing through unit area in unit time. When multiplied by the density this obviously gives the flux ($\rho q\,\mathrm{kg\,m^{-2}s^{-1}}$) and in the absence of sources and sinks, conservation of mass within a volume (10.5) must apply.

$$\frac{\partial\rho}{\partial t} + \nabla\cdot(\rho\underline{q}) = 0. \tag{10.45}$$

Assuming that the density is constant or varies only slowly with time, then setting the first term in (10.45) to zero and using Darcy's equation to replace $\underline{q}$ by $\nabla\phi$

$$\nabla\cdot\nabla(\phi) = \nabla^2\phi = \frac{\partial^2\phi}{\partial x^2} + \frac{\partial^2\phi}{\partial y^2} + \frac{\partial^2\phi}{\partial z^2} = 0. \tag{10.46}$$

Equation (10.46) is Laplace's equation as applied to the spatial variation of hydraulic potential within the volume considered.

10.6.2 *Unconfined aquifers: Dupuit approximation*

The horizontal components of discharge, q_x and q_y, are generally much greater than the vertical component (except near sources and sinks such as wells or drains) in which case the hydraulic potential, ϕ, is essentially independent of z. The assumption that the vertical component of groundwater flow may be neglected is known as the Dupuit approximation. Clearly, the assumption does not hold very well for situations where there are significant vertical components of discharge, but it does apply to unconfined aquifers.

An unconfined aquifer is a water bearing layer located above an impervious layer, and where the top is open to the atmosphere through the pores of the unsaturated zone. The situation is illustrated schematically in Fig. 10.7. We take the interface between the water bearing and impervious layers as $z = 0$ and consider the discharge though an elemental strip of height $h = \phi$ perpendicular to the direction of ground water flow.

The mass flow rate through the strip is given by the product $\rho h \underline{q} = \rho \phi \underline{q} (\text{kg s}^{-1})$, and making use of the Darcy equation:

$$\rho \phi \underline{q} = -\rho \phi K \nabla \phi. \qquad (10.47)$$

In the steady state and for constant density, conservation of mass requires that

$$\nabla \cdot (\phi \underline{q}) = -K \nabla \cdot (\phi \nabla \phi) = 0. \qquad (10.48)$$

In the Dupuit approximation, $q_z = 0$ and if the hydraulic conductivity is a constant, equation (10.48) reduces to the two dimensional Laplace equation in ϕ^2:

$$\frac{\partial}{\partial x} \left(\phi \frac{\partial \phi}{\partial x} \right) + \frac{\partial}{\partial y} \left(\phi \frac{\partial \phi}{\partial y} \right) = \frac{\partial^2 \phi^2}{\partial x^2} + \frac{\partial^2 \phi^2}{\partial y^2} = 0. \qquad (10.49)^{\text{h}}$$

Equation (10.49) differs from (10.46) in that it refers to the total mass flow rate through a vertical section and includes variations in ϕ with x and y.

Vertical discharge is seldom totally zero, for example, evaporation and capillary action will provide upward discharge, while following rain there will be a downward discharge as the rain water percolates down through the top soil. Equation (10.49) explicitly assumes that $q_z = 0$ and where this is not the case, then the more general equation (10.46) should be used.

As a simple example of an unconfined aquifer, consider the one-dimensional case of an irrigation canal connected to the neighbouring field (the unconfined aquifer) through an earth bank. At some distance x_1 from the edge of the canal ($x = 0$), the field will have an equilibrium groundwater head ϕ_1. The groundwater head in the edge of the earth bank will just be the water level in the canal (ϕ_0) as depicted in Fig. 10.8. The problem is to determine the variation in ϕ with distance from the canal with the given boundary conditions.

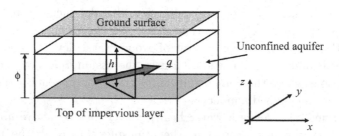

Fig. 10.8   Unconfined aquifer.

---

^{h}Note $\frac{\partial^2 \phi^2}{\partial x^2} \equiv \frac{\partial}{\partial x} \left(\frac{\partial \phi^2}{\partial x} \right) = \frac{\partial}{\partial x} \left(2\phi \frac{\partial \phi}{\partial x} \right)$; similarly for $\frac{\partial^2 \phi^2}{\partial y^2}$. The common factor of 2 divides out.

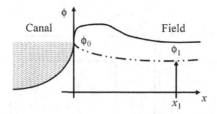

Fig. 10.9   Variation of ϕ with distance from a canal.

Integrating (10.49) twice and applying the boundary conditions:

$$\phi^2(x) = \phi_0^2 - (\phi_0^2 - \phi_1^2)\frac{x}{x_1}. \tag{10.50}$$

The groundwater head varies parabolically with distance from the canal edge. As a result, there will be a discharge through the canal banks to the surrounding field. The discharge per unit of length of canal bank at x is

$$\phi q = -K\phi\frac{\partial\phi}{\partial x} = -\frac{1}{2}K\frac{\partial\phi^2}{\partial x} = (\phi_0^2 - \phi_1^2)\frac{K}{x_1}. \tag{10.51}$$

In this example, the canal is a constant source maintaining the groundwater head in the banks and is assumed to be sufficient to maintain any discharge through the bank to the field. The level in the field will be subject to loss due to transpiration in the plants, and, providing there is no rain an approximate dynamic equilibrium should be established. Under these conditions, the discharge rate should be independent of x as indicated in equation (10.51).

10.6.3 *Wells in unconfined aquifers*

If the saturated surface intersects the surface, e.g. on a hillside, there will be a natural outflow which can be collected, otherwise a well shaft must be sunk into the aquifer to extract water. The well is generally drilled completely through the aquifer to the impermeable layer and lined with a perforated casing, allowing water to enter while maintaining the structural integrity of the shaft.[2]

When a well is first sunk through the saturated zone, it will fill with water up to the level of the phreatic surface as discussed earlier. However, once water is withdrawn from the well, its level will decrease because discharge from the surrounding medium will be insufficient to replenish the loss. There will be a corresponding decrease in the local head as illustrated in Fig. 10.10. At radial distances far from the well, the head will be undisturbed at the original level, h_o, but closer to the well the head will be progressively reduced in a 'cone of depression' culminating in the new head in the well, h_w.

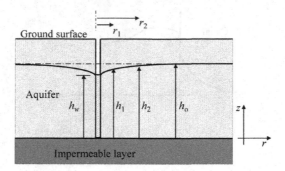

Fig. 10.10   Schematic diagram of a well in an unconfined aquifer.

The discharge towards the well is given by Darcy's equation. In the steady state and assuming cylindrical symmetry:

$$q(r) = -K\frac{d\phi}{dr} \tag{10.52}$$

When the well is drilled completely through the aquifer to the impermeable region, then the total discharge through a concentric cylindrical shell at r must be given by:

$$Q = 2\pi rzq = -2\pi rzK\frac{dh}{dr}. \tag{10.53}$$

where (dh/dr) is the water head gradient at r. Integrating between h_1, r_1 and h_2, r_2:

$$Q\int_{r_1}^{r_2} \frac{dr}{r} = 2\pi K \int_{h_1}^{h_2} z\,dz \tag{10.54}$$

gives the total discharge from r_2 to r_1:

$$Q = \frac{\pi K(h_2^2 - h_1^2)}{\ln(r_2/r_1)}. \tag{10.55a}$$

The lowering of the water head is known as the drawdown, and if this is not too large, (10.55a) is a reasonable approximation to the steady state discharge into the well, which may be estimated by replacing r_1 by r_w, the radius of the well and r_2 by r_0, a sufficiently large radial distance such that $h = h_0$.

$$Q_{\text{Well}} = \frac{\pi K(h_0^2 - h_w^2)}{\ln(r_0/r_w)}. \tag{10.55b}$$

10.6.4 *Wells in confined aquifers*

In confined aquifers, the saturated water bearing layer is sandwiched between two impervious layers. The lower water bearing stratum is no longer in direct contact with the atmosphere and is in addition under pressure from the weight of the

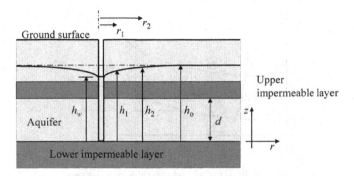

Fig. 10.11   Well in a confined aquifer.

overlying impervious level. In these circumstances, ϕ (taking the bottom of the water bearing layer as $z = 0$) will lie above the top of the saturated layer, i.e. the pore pressure at the top of the aquifer will be greater than atmospheric pressure. Figure 10.12 shows a schematic diagram of a well in a confined aquifer of constant thickness, d.

The steady radial discharge into the well from the surrounding is obtained in the same way as for the unconfined aquifer. The principle difference is that the only flow into the well comes from the saturated layer of thickness, d. Equation (10.53) must then be rewritten:

$$Q = 2\pi dK \frac{dz}{dr}. \tag{10.56}$$

Integrating as before from h_1, r_1 to h_2, r_2:

$$Q = 2\pi dK \frac{(h_2 - h_1)}{\ln(r_2/r_1)}. \tag{10.57}$$

10.6.5 *Equipotentials and stream lines*

In the case of wells, whether in confined or unconfined aquifers, the assumption that there is no vertical discharge is not really tenable. In any practical well from which water is being drawn, replenishment from the surroundings results in a significant gradient in the local water head which will affect the discharge into the well, and a more general approach is required.

Many of the problems of ground water flow can be treated as two dimensional, and it is common to treat them as such. The hydraulic potential, ϕ, is considered as a two-dimensional potential, $\phi(x,y)$[i]; lines of $\phi =$ constant are therefore equipotential

---

[i]For simplicity we shall assume $K = 1$.

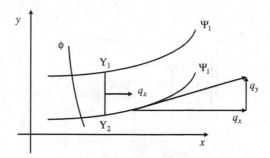

Fig. 10.12   Diagram illustrating the relationships between q, ϕ and ψ.

curves. By definition, their total differential must vanish:

$$d\phi = \frac{\partial \phi}{\partial x}dx + \frac{\partial \phi}{\partial y}dy = 0. \tag{10.58}$$

For two-dimensional steady $\partial q/\partial t = 0$ incompressible flow, the equation of continuity may be written:

$$\frac{\partial q_x}{\partial x} + \frac{\partial q_y}{\partial y} = 0. \tag{10.59}$$

It can be shown that since (10.59) is true for all (x, y) there exists a function known as the 'stream function', $\psi(x,y)$,[1] defined as:

$$q_x = \frac{\partial \psi}{\partial y}; \quad q_y = -\frac{\partial \psi}{\partial x}. \tag{10.60}$$

Substituting for q_x and q_y from Darcy's equation (10.44):

$$\frac{\partial \phi}{\partial x} = \frac{\partial \psi}{\partial y} \quad \text{and} \quad \frac{\partial \phi}{\partial y} = -\frac{\partial \psi}{\partial x}. \tag{10.61}$$

Equations (10.61) are the 'Cauchy-Riemann' equations.

As before, the total differential of ψ is zero for $\psi = $ constant:

$$d\psi = \frac{\partial \psi}{\partial x}dx + \frac{\partial \psi}{\partial y}dy = 0 \quad \text{or} \quad \frac{\partial \psi}{\partial x}dx = -\frac{\partial \psi}{\partial y}dy. \tag{10.62}$$

Figure 10.12 shows a pair of stream lines plotted on the xy plane. The slope of say, ψ_1 is, using (10.62) and (10.60):

$$\frac{dy}{dx} = -\left(\frac{\partial \psi}{\partial x}\right) \bigg/ \left(\frac{\partial \psi}{\partial y}\right) = \frac{q_y}{q_x}. \tag{10.63}$$

Figure 10.12 demonstrates that the stream lines are parallel to the discharge vector.

The slope of the equipotential ϕ = constant is from the total differential (10.58).

$$\frac{dy}{dx} = -\left(\frac{\partial \phi}{\partial y}\right) \bigg/ \left(\frac{\partial \phi}{\partial x}\right) = -\frac{q_x}{q_y} \tag{10.64}$$

which is normal to the stream line.

The specific discharge vector is everywhere *perpendicular* to the equipotentials and *parallel* to the stream lines. Flow along a given stream line is by definition constant, but it also follows that the discharge between neighbouring stream lines will be constant; the magnitude being determined by their spacing. To illustrate this, consider the total discharge through the section Y_1Y_2 of unit thickness in Fig. 10.12.

$$Q_{Y_1Y_2} = \int_{\psi_1}^{\psi_2} q_x dy = \int_{\psi_1}^{\psi_2} d\psi = (\psi_2 - \psi_1). \tag{10.65}$$

The total discharge is dependent only on the difference in ψ. The spacing between the stream lines gives the relative magnitudes of the flow velocities; the closer together they are the higher the velocity.

Stream lines must start at a source (e.g. a river) or a water surface and end at a sink (e.g. a well) or at another water surface. In the case of the wells discussed in the previous two sections, the topology of land was flat, and flow assumed to be horizontal. In this rather idealised system, stream lines terminate at the well, and radiate outwards isotropically towards the source(s), while the equipotentials will form a series of concentric circles.

A more general case is where the well is located within some flow field[3] as illustrated in Fig. 10.13. Here there is a uniform flow in the negative x-direction

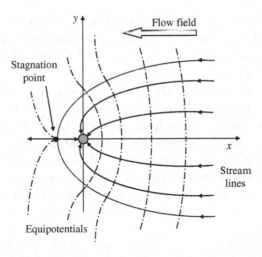

Fig. 10.13   Distribution of stream lines and equipotentials around a well in a uniform flow.

and the presence of the well distorts the stream lines and equipotentials. To the left of the well (which for convenience has been situated at the origin) there will be a point where the discharge towards the well is just equal to the uniform flow. This is termed the stagnation point, and the stream line passing through it is the limiting or stagnation flow line. This is still a somewhat idealised case, and although the equipotentials and stream lines are not distributed isotropically, they are symmetrical about the x axis, i.e. the flow direction. Real situations (which do not conform to simple geometries), require modelling on a case by case basis. The stream line-equipotential approach is particularly useful as it provides a network, which may be used as the basis for finite element analysis, an effective method for dealing with these problems.

10.7 Problems

1. Show that equation (10.13) is a solution to the transport equation (10.12).

   A small drop of water is inadvertently released into a space capsule where it evaporates. A long time later, the concentration at a distance of 0.5 m is found to be 1% that of the concentration at the point of release at the same instant of time. Estimate the corresponding value of the standard deviation, σ. Estimate the corresponding concentration at a distance of 0.25 m.

2. A cylindrically shaped pond, radius a is bounded by an earth wall outer radius b. If the hydraulic potential inside the wall is ϕ_a and outside the wall it is ϕ_b show that:

$$\phi^2(r) = \phi_a^2 - \frac{\phi_a^2 - \phi_b^2}{\ln(b) - \ln(a)}(\ln(a) - \ln(r))$$

   where $\phi(r)$ is the hydraulic potential at some radius r ($a < r < b$). Assume axial symmetry.

   If the hydraulic conductivity of the earth wall is K, obtain the *total* discharge Q from the pond in terms of the *average* hydraulic potential and the *difference* in hydraulic potential between the inside and outside of the earth wall, and show that Q is independent of the radius.

3. A cylindrical pond 6 m in diameter is surrounded by an earth wall 1 m thick. If the hydraulic conductivity of the wall is 10^{-8} ms^{-1}, the difference in hydraulic potential is 2 m and the mean hydraulic potential is 3 m, estimate the total seepage from the pond over the course of a year, assuming no changes in the conditions. If the pond was contaminated with some pollutant at a concentration of 100 p.p.m., calculate the volume of the contaminant that leaks out of the pond over the year assuming that the contaminant has the same hydraulic conductivity.

$$\left(\text{N.B. in cylindrical polar co-ordinates: } \nabla^2 = \frac{1}{r}\frac{\partial}{\partial r}\left(r\frac{\partial}{\partial r}\right) + \frac{1}{r^2}\frac{\partial^2}{\partial\theta^2} + \frac{\partial^2}{\partial z^2}\right)$$

References

1. For a much more complete treatment of dispersal of pollutants in rivers, see Egbert Boeker and Rienk van Grondelle, *Environmental Physics*, 2nd edn. (John Wiley & Sons, 1999).
2. Viessman Jr. and G. L. Lewis, *Introduction to Hydrology*, 4th edn. (Harper & Collins, 1995).
3. Karlheinz Spitz and Joanna Moreno, *A Practical Guide to Groundwater and Solute Transport Modelling* (John Wiley & Sons, 1996).

Chapter 11

STEWARDING THE ENVIRONMENT

11.1 Introduction

The primary purpose of this book was to discuss some of the physics that govern the environment, and the human activities that have a direct bearing on that environment, notably the generation of electrical power and transportation. Nevertheless, no text on environmental physics can completely ignore questions about the use and abuse of the Earth's resources, and physical scientists have a unique contribution to make to current debates over climate change, consumption, pollution etc. It is incumbent on the profession to undertake 'good science' that seeks to inform the debate, sound warnings where necessary and lay out the options.

Both material and terrestrial energy reserves are finite and the excessive and wasteful exploitation of these will obviously have significant and potentially disastrous consequences for the future. The notion that resources are limited is not new. It was probably first brought to the public domain by the Club of Rome report 'Limits to Growth' in 1972.[1] This study attempted to model patterns of consumption and resource into the future and although the models it employed have since been shown to be overly simplistic, the report did alert the wider community to the problem.

Similarly, the reckless or merely thoughtless discard of waste products into the environment can have equally devastating effects. Although the emission of greenhouse gases and global warming have become the dominant issue, disposal of waste in landfill and/or in the seas and oceans are also matters of concern.

The focus of this final chapter is less on the physical principles that have formed the main content of this book. Rather, the intention here is to highlight some of the less tangible considerations; the socio-political and philosophical framework which influence the thinking of policy makers, the uncertainties and unknowns that complicate priorities and the pressures from conflicting constituencies each promoting their own particular point of view.

11.2 Technical Solutions and Public Perceptions

We currently live in the Period of Enlightenment, where perceptions are based on a 'scientific' model of the world. It dates from the mid-17th century when the notion that the universe was capable of rational description and understanding gradually became accepted. Scientific discovery replaced superstition and in due course led to technological invention in a process that has gathered pace to the extent that science and technology have become almost indivisible. New science finds applications almost immediately and the potential technological benefits compel governments to invest significant proportions of their country's wealth in scientific infrastructure and education. There are at least two unfortunate consequences that stem from this perspective.

The successful application of scientific principles in solving problems has generated the idea that there will always be a 'technical fix' for any problem — some solutions can invariably be found if sufficient effort is directed at the problem. It has indeed been the case that technical fixes have often been found. However, there is no guarantee of this and the belief that all problems are amenable to technological solution leads to complacency. There may not be such a solution, or perhaps not one that is affordable or that can be developed in sufficient time. Moreover, technical solutions have not always worked or else have had unintended and unforeseen consequences. Sometimes other approaches must be used.

The trend towards specialisation, particularly in the physical sciences, have made the subjects increasingly incomprehensible even to fellow practitioners not directly in the field. As a result, there is often distrust of scientific opinion which perforce has to be couched in terms of probability, jargon and mathematical abstraction. The distrust is compounded when that opinion is perceived (rightly or wrongly) to be too closely allied to some corporate or vested interest, be it industrial, governmental or environmental. Thus while an overly naïve belief that a technical solution exists may breed complacency, an exaggerated scepticism may prejudice a perfectly reasonable and effective technological solution.

11.3 Future Energy Supplies

Projections of future energy needs in the West universally predict a shortfall in electricity generating capacity. The situation is being compounded in many countries as aging nuclear facilities are taken out of service, having for the most part far exceeded their design lifetimes. There are a variety of technical options for meeting the anticipated energy deficit, but there are drawbacks associated with each one.

The easiest way to meet the shortfall would be to simply build more fossil-fuelled stations. Gas-fired stations are well developed, relatively clean and can be constructed quickly. However, the global reserves of gas are diminishing rapidly and many formerly self-sufficient countries including the UK are no longer so, and are having to import gas in a highly competitive global market controlled by supply. This poses strategic dilemmas for governments and the prospect that prices will rise inexorably. Some countries have extensive coal reserves and a further option is to resurrect coal burning power stations. Technological developments can be expected to improve efficiencies and significantly limit the sulphur and particulate emissions that plagued earlier designs. Undoubtedly, much of the energy gap could be met by the construction of new coal and gas-fired power stations, but only at some cost to the environment, and at best fossil-fuelled stations can only be a short term solution. However, if the CO_2 emissions can be successfully sequestered, fossil-fuelled stations may provide interim power while alternative supplies and/or infrastructures are developed and installed.

On a longer time scale, future energy needs could in principle be met by the increased use of nuclear power. The current generation of nuclear reactors are significantly more efficient than earlier designs and this is of course a carbon free technology. Nuclear power facilities fit readily into the existing grid structure, providing substantial inputs of electrical power at network nodes. Integration into the existing infrastructure would be straightforward and comparatively inexpensive, although the plant itself is not cheap; the shielding requirements alone adding considerably to the cost. Nuclear power consumes the comparatively rare isotope of uranium ^{235}U, like coal and gas, also a finite resource. The major objection to nuclear power is the long-lived radioactive waste. While technologies exist for its dispersal (i.e. in stable glasses) within the environment, the timescales are so long that they may well outweigh the advantages of zero carbon emission. In addition, there is the not inconsiderable cost of dismantling the station at the end of its useful working life. Perhaps the greatest impediment to the adoption of nuclear power is the mistrust of the wider public. Objections to the building of a new station can delay the start of construction by many years. That in itself would render this a long-term solution, if indeed a solution it is.

Much store is set by the implementation of renewable sources of energy, though these too are not without drawbacks. They are not well suited to large scale grid systems since they can supply only a modest amount of power at any given location. Moreover, the power input is uncontrolled, depending in the case of wind farms on the vagaries of the weather and for solar power sources, on sufficient daylight. Renewable sources, it is argued, are much more compatible with distributed systems where power is generated close to the point of use, cutting transmission losses and effectively offsetting much of the initial setup costs. Intrinsic to this argument is the assumption that the distributed systems are preferable, a premise that may not be altogether well founded. For example, what is a community reliant on say wind power to do when there is either too little or too much wind for the turbines to function? How are major energy intensive facilities, whether social (hospitals, schools etc.) or industrial (on which the livelihoods of many depend) to be supplied reliably? There are very good reasons why the electricity grid system has developed over many decades in the way it has.

Some places are of course fortunate in that local circumstances permit a wider range of choices. Iceland is a good example. Blessed with an abundance of geothermal energy, it is not only able to generate an adequate supply of electricity, but has surplus energy available to electrolyse water (with which it is also well endowed) and so producing hydrogen. The country plans to move to an all hydrogen economy in which all transport, including its fishing fleets, will be hydrogen powered, and even aspires to be a net exporter of energy.

One obvious solution is simply to consume less energy. Savings made by improved house insulation can reduce both heat loss in colder climates and heat input in hotter places. Double glazing, cavity walls and better building materials would reduce the demand for additional heating in the one case and air-conditioning in the other. This course of action is relatively straightforward to introduce for new-build by legislation, but rather less simple to 'retrofit' to existing homes. Householders may be unwilling or unable to meet the capital expenditure required, and would probably require financial incentives (e.g. grants) beyond the savings made in the regular gas or electric bills.

The options available to decision makers are complex and for the most part opaque. The majority of developed countries are facing energy shortages on timescales that allow little prevarication. There are advantages and disadvantages associated with all the options and it may well prove to be the case that the best course of action is an integrated combination of them all. They are largely technical solutions designed to cope with the anticipated shortfall, the underlying cause of which is the propensity to increase energy requirements seemingly without limit. Moderation of lifestyles in the developed world could result in a significant reduction of the energy demand. In a large measure this is a question of public education.

11.4 Pollution Issues

Closely allied to the generation of electrical power are questions of pollution. This tends to focus on greenhouse gases or in the case of nuclear power, radioactive waste. All hydrocarbon fuels ultimately produce CO_2, and therefore contribute to the greenhouse load imposed on the atmosphere, coincidentally making it more difficult to meet Kyoto quotas for signatory nations. As discussed earlier (Sec. 6.7), methods do exist and are being further developed for the sequestration of CO_2, for example by pumping the gas into deep geological formations and (controversially) into the bottom of deep seas. Where this can be used to release inaccessible oil and methane, there are additional benefits. Proponents of sequestration technologies point to the potentially vast CO_2 storage capacity available, however the efficacy of these solutions has yet to be fully demonstrated. They are, in addition, only practical in dealing with the emissions from large installations.

Vehicle emissions of CO_2 cannot be removed in this way implying that alternatives will have to be found if private and public modes of transport are to be sustained. Currently, the most promising options appear to be based on hydrogen (Sec. 9.4), where combustion either directly in internal combustion engines or in fuel cell based systems eliminate CO_2 altogether. However, hydrogen propulsion is not without its own problems, not least regarding the twin issues of generation and portability. As a secondary fuel source, hydrogen must be produced by the consumption of some other fuel. Clearly, this only makes sense if the primary source is either renewable or carbon neutral. The transition to a hydrogen-based economy would be enormously complicated and expensive. It will entail the establishment

of an entirely new distribution infrastructure that for a significant period of time
will have to operate in parallel with the existing petroleum and diesel supply
system. Nevertheless, all the major car manufacturers have advanced research
and development programmes on the expectation that hydrogen powered vehicles
are inevitable, and some public transport operators are already beginning to use
hydrogen.

In the meantime, there is a need to reduce the dependence on petrol and diesel
fuels for road vehicles. In Western Europe, fiscal strategies (i.e. taxation) have
been used to discourage the use of fuel inefficient and correspondingly high carbon
emitting cars. As a result, cars are becoming increasingly efficient and less polluting.
There has been a tradition of high fuel taxes in the countries of the EU and this
has no doubt contributed to a general acquiescence of the principle and has ensured
that fuel efficiency is a high priority when purchasing a new car. There are, however,
limits to what individuals will tolerate and economies can accommodate, while in
some countries such as the USA there is considerable opposition to the principle
of taxation *per se*. However, even in the USA there is a growing recognition that
consumption will have to be restrained.

11.5 Uncertainty and Risk

The choices that face decision makers are invariably plagued by uncertainty. Is there
a technological option? If so, what will it cost and over what timescale? Can these
questions even be answered at the time the decision has to be made? The answers
are seldom straightforward. As an analogy, will straightening a particular stretch
of road improve safety and so reduce accident rates or will it lead to an increased
level of traffic thereby negating the intended advantage?

Risk is a number that signifies the probability that a given adverse event will
occur within some specified period of time. *Every* activity has some risk associated
with it; the question is what is acceptable. There is obviously no correct or objective
answer. An individual may accept a high risk to pursue a hazardous recreational
activity, e.g. caving or parachuting with equanimity, but object strongly to a very
much lower level of risk from some neighbouring facility such as a power station.

In the implementation of some policy, particularly involving infrastructure —
the building of new road or rail links, building a new power station, locating
industrial plant — a government or planning authority has to take risk into account.
An estimate must be made of the *increase* in risk to the local population of the new
installation. This is usually amenable to fairly precise calculation, aided if necessary
by sophisticated computer modelling. The question of whether the additional risk
to the neighbourhood is acceptable is primarily a political one. The facility may
increase employment opportunities, attract additional inward investment boosting
the local economy with a resulting enhancement to the health and well-being of
many in the community. In all probability, not everyone in the locality will benefit
and some may be strongly opposed in principle. For example, wind farms frequently

generate implacable opposition because they are felt to be visually obtrusive, especially as they tend to be located on remote and scenic country sites where mean wind speeds will be high. The additional risk to anyone in the local community, on the other hand, is probably negligible.

The precision with which the risk can be stated depends on the experience of the hazard in question. Thus the risk of a road traffic accident is well established, that of a significant nuclear incident (fortunately) rather less so. Road traffic accidents are much more common than are nuclear disasters. However, the devastation produced by a single nuclear accident will greatly exceed that associated with a single road accident. Any analysis of cost-benefit must therefore include the harm that would result, should the event ever occur. This gives rise to the idea of detriment which is the product of the risk of an event and the ensuing harm. Thus the risk of a nuclear incident may be very small (in fact, there have only been a handful world-wide), the harm substantial, but the detriment is (or was) deemed to be acceptable. The example of nuclear power also serves as a warning that timescales should be included in the calculation. The risk to a community from a nuclear power station persists in the form of radioactive waste long after the station has been taken out of commission and dismantled.

11.6 Choices and Stewardship

In a free society, decisions made on behalf of the population at large must be accepted by them. A government cannot successfully implement a given course of action, however sensible, unless the country as a whole is prepared to follow suit. It may discourage the use of cars, but probably can never prevent it and even indirect fiscal methods such as increased taxation may have only limited effect. Nevertheless, choices have to be made and leaders must seek to explain their decisions and engage the public's support if the selected course of action is to succeed. This will probably mean both persuasion and possibly 'coercion' through the framing of appropriate new laws in the country's legislature.

The current economic climate is based on the stimulation of consumption and demand by the provision of evermore new and exciting 'must have' products that differ from the previous model in only the trivial details. Obsolescence is no longer built into the physical product as was once the case, indeed quality and reliability is generally of a very high standard. Rather 'obsolescence' is built into public perception, and consumer durables are frequently discarded before the end of their design life. This is clearly not sustainable. The principle should be that every resource consumed should reflect its replacement value and this must include the cost incurred in the disposal of any waste. For every tree chopped down, another should be planted.

The concept of stewardship is that the "steward" is responsible for the care and maintenance of the "master's" property. The steward did not own the property, although it was at his/her disposal on the understanding that it was to be used responsibly. Those with a religious faith see the world as belonging to God and only entrusted to humanity for the wider benefit of all. It is an ethical viewpoint, at odds with accepted norms of ownership. The principle that everyone, regardless of religious persuasion, share a responsibility for the upkeep and maintenance of the planet and its environment is gradually gaining wider acceptance. Some promote it as a new paradigm shift that sets the 'old' western civilisation against a 'new' one based on sustainability. A new paradigm or not, it is the case that difficult political decisions — on global warming, consumption of resources, pollution etc. — will have to be taken against a background of uncertainty and in the face of vested interests.

Reference

1. Donnella H. Meadows, *Limits to Growth*, Report to the Club of Rome (1972), ISBN 0330241699.

INDEX